Communication Ethics and Crisis

Communication Ethics and Crisis

Negotiating Differences in Public and Private Spheres

S. Alyssa Groom and
J. M. H. Fritz, editors

FAIRLEIGH DICKINSON UNIVERSITY PRESS
Madison, NJ

Published by Fairleigh Dickinson University Press
Co-Published with The Rowman & Littlefield Publishing Group, Inc.
4501 Forbes Boulevard, Suite 200, Lanham, Maryland 20706
www.rowman.com

Estover Road, Plymouth PL6 7PY, United Kingdom

British Library Cataloguing in Publication Information Available

Library of Congress Cataloging-in-Publication Data

Communication ethics and crisis : negotiating differences in public and private spheres / [edited by] S. Alyssa Groom and Janie Harden Fritz.
p. cm.
ISBN 978-1-61147-449-7 (cloth : alk. paper) — ISBN 978-1-61147-450-3 (electronic)
1. Communication—Moral and ethical aspects. 2. Risk communication—Moral and ethical aspects. 3. Communication in organizations—Moral and ethical aspects. 4. Crisis management. 5. Emergency management. I. Groom, S. Alyssa, 1975– II. Fritz, Janie M. Harden, 1961–
P94.C5696 2012
175—dc23 2011030593

∞™ The paper used in this publication meets the minimum requirements of American National Standard for Information Sciences—Permanence of Paper for Printed Library Materials, ANSI/NISO Z39.48-1992.

Printed in the United States of America

Contents

Acknowledgments

The editors would like to thank Kristen Jenkins , G. Blake Plavchak, and Dan Snyder for their able assistance in the preparation of this volume.

Contributors

Ronald C. Arnett (Ph.D., Ohio University) is chair and professor in the Department of Communication and Rhetorical Studies, Duquesne University, specializing in philosophy of communication and communication ethics. In addition to over 60 articles and three edited books, his works include *Communication Ethics Literacy: Dialogue and Difference* (2009, with Janie M. Harden Fritz and Leeanne M. Bell); *Dialogic Confession: Bonhoeffer's Rhetoric of Responsibility* (2005); *Dialogic Civility in a Cynical Age: Community, Hope, and Interpersonal Relationships* (1999, with Pat Arneson); *Dialogic Education: Conversations about Ideas and Between Persons* (1992); *Communication and Community: Implications of Martin Buber's Dialogue* (1986); and *Dwell in Peace: Applying Nonviolence to Everyday Relationships* (1980). Email: arnett@duq.edu

Janie M. Harden Fritz (Ph.D., University of Wisconsin-Madison) is associate professor of communication and rhetorical studies at Duquesne University. Her research examines communicative practices that constitute, sever, and restore the ties that bind individuals to institutions. She coedited (with Becky L. Omdahl) *Problematic Relationships in the Workplace* (Peter Lang, 2006) and coauthored (with Ronald C. Arnett and Leeanne M. Bell) *Communication Ethics Literacy: Dialogue and Difference* (Sage, 2009). Her work has appeared in several journals, including *Management Communication Quarterly*, *Journal of Business Ethics*, *Communication Research Reports*, *Journal of Public Management and Social Policy*, *Communication Monographs*, and *Journal of Business Communication*. Email:harden@duq.edu

Pat J. Gehrke (Ph.D., Pennsylvania State University) is associate professor in the Program in Speech Communication & Rhetoric and the Department of English at the University of South Carolina. His work focuses on the history of rhetoric and communication in the United States and the role of ethics and politics in rhetorical theory. He brought these two interests together in his latest book, *The Ethics and Politics of Speech* (Southern Illinois University Press 2009). His writings have also appeared in numerous journals, such as *The Quarterly Journal of Speech*, *Philosophy and Rhetoric*, *Philosophy Today*, and *Critical Studies in Media Communication*.

S. Alyssa Groom (Ph.D., Duquesne University) is assistant professor of communication and rhetorical studies at Duquesne University. She does applied and theoretical scholarship in integrated marketing communication and has published two book chapters: "Engaging the Rhetorical Consciousness of an Organization for Dynamic Communicative Exchange and Advertising: Storytelling in the Marketplace of the 21st century," and "Identity and Social Economy" for the *Encyclopedia of Identity*. In 2008, she published "Integrated Marketing Communication: Anticipating the Age of Engage" for the journal *Communication Research Trends*. Email: grooms@duq.edu

Craig E. Mattson (Ph.D., Regent University), is professor of communication arts at Trinity Christian College. His research centers on three main areas: public theology, public intellectualism, and the rhetoric of fiction. His research has been published in journals such as the *Journal of Communication and Religion*, *Tradition and Discovery*, *Christianity and Literature*, *Christian Scholar's Review*, and *Books and Culture*. Email: craig.mattson@trnty.edu.

John H. Prellwitz (Ph.D., Duquesne University) is assistant professor of communication at the University of Pittsburgh at Greensburg. His doctorate includes emphases in public communication/integrated marketing and ethics. His dissertation, "Nietzsche's Dialogic Ethic After Illusion: Rhetoric and Difference," was nominated for Gerald R. Miller Outstanding Doctoral Dissertation Award sponsored by the National Communication Association. His current research interests include examination of the implications of Friedrich Nietzsche's thought for rhetoric, philosophy of communication, and communication ethics; and intersections between philosophy of communication, communication ethics, film, propaganda, and media ecology. Email: jhp15@pitt.edu.

Deanna M. Sellnow (Ph.D., University of North Dakota), is the Gifford Blyton Endowed Professor and director of undergraduate studies in the Department of Communication at the University of Kentucky. Her areas of specialty include instructional communication, educational assessment, popular culture rhetoric (particularly popular music), and gender communication. She has published extensively in the area of instructional communication. Her work has appeared in a number of journals, including *Communication Research Reports, Communication Education, Critical Studies in Media Communication, Women's Studies in Communication,* and *Basic Communication Course Annual.* E-mail: Deanna.Sellnow@uky.edu.

Timothy L. Sellnow (Ph.D., Wayne State University) is professor and associate dean at the University of Kentucky. His primary research and teaching focus is on risk and crisis communication. Much of his recent research focuses on strategic communication for mitigating the impact of and maintaining resilience in response to potential terrorist attacks on the United States. He currently serves as theme leader for the risk communication research team at the National Center for Food Protection and Defense, a Center of Excellence sponsored by the Department of Homeland Security, and has served as a risk and crisis communication consultant for the Centers for Disease Control and Prevention. E-mail: Tim.Sellnow@uky.edu.

Celeste Grayson Seymour (M.A., Duquesne University) is a Ph.D. student at Duquesne University. Her research interests include philosophy of communication, existential phenomenology, and communication ethics. She has co-authored work featured in *Communication Research Trends* and contributed essays to the *Encyclopedia of Identity,* edited by Ronald L. Jackson III. Email: celeste.grayson@gmail.com.

Steven J. Venette (Ph.D., North Dakota State University) is assistant professor of speech communication at the University of Southern Mississippi. His teaching and research interests include risk and crisis communication, organizational communication, and argumentation. Recent projects include work for the Department of Homeland Security's National Center for Food Protection and Defense, the U.S. Centers for Disease Control and Prevention, USDA Plant Protection and Quarantine, and USDA Cooperative State Research, Education, and Extension Service. E-mail: steven.venette@usm.edu.

Introduction

Communication Ethics and Crisis: Negotiating Differences in Public and Private Spheres

S. Alyssa Groom and J. M. H. Fritz
Duquesne University

Communication ethics and crisis communication are rapidly growing areas of theoretical and applied exploration. In spite of their inherent differences, they intersect—often collide—every day in public and private contexts that shape marketplace, interpersonal, and organizational communication practices and outcomes. Family, collegial relationships, organizations and the very nature of the human condition identify the tangible spaces in which this relationship between communication ethics and crisis communication emerges. Although they are not always explicitly connected or discussed, lived experience makes it clear that the two are inextricably linked.

In hopes of forging broader pathways into the experience and response of crisis situations, this edited work presents ideas that offer constructive engagement of the interrelation of communication ethics and crisis communication beyond the organizational context. It begins from a point of convergence known as "crisis," or anywhere the relationship between a lived experience is met with discordance that rends concordance, thus calling for an experience where "concordance meets discordance" in an attempt to address the situation (Ricoeur 1984, 31). In such moments of rending, communication ethics comes to the forefront, understood as a philosophy of communication embracing an understanding of "the good" that emerges in interaction with others and with a given historical moment, requiring dialogic negotiation to discern potential responses among competing goods (Arnett, Fritz, and Bell 2009). In this situation, meeting discordance produces a rhetorical litmus test as people and/or organizations grapple with a problem, evaluation, context, set of goods, and response as a means of working through the crisis. Some do work through it. Strong and resilient, they are able to meet the situation with the hope and possibility of landing on solid ground. They can experience the catharsis of Ricoeur's (1984) discordant concordance and recover. They are alkaline. Others are set hopelessly afloat by crisis, unable to meet the situation at hand. Instead of asking essential questions, such as "How should multiple goods be engaged in order to make the best decision in a given moment?" they are vexed with panic, aversion, emotional myopia, and other problematic indicators of crisis. There is no constructive engagement of those indicators towards a hopeful outcome, no resolution, and perhaps no attempt to survive. They are acidic.

The contributors to this volume offer a collection in which ethics and crisis move toward alkaline association with one another; they form a conversation about ideas aimed at moving the two areas into a stronger and more overt relationship. In the spirit of Aristotle (1954), each author contributes to an understanding of the intersection between communication ethics and crisis communication as a rhetorical encounter that necessitates the engagement of all available means of persuasion in order to address the situation with the best hope of a constructive outcome. As a teacher of rhetoric, Aristotle advanced practical wisdom, phronesis, as a foundational aspect of action towards an ethical and good end. Using practical wisdom to discern "all available" means of persuasion required attention to more than just the moment. This is especially true when faced with a crisis. Practical wisdom must be careful to consider the past, present, and future. It must be willing to gain insight from traditional culture as well as from voices present in the room. It must ask questions regarding the appropriateness of a given choice, the effect of a decision, and the implications of action on multiple audiences, including those who have come before and those who will follow.

Practical wisdom, in the form of each essay that follows, offers perspectives on reflection and response-ability as guides for ethical decision making in crisis situations. While each begins from a different starting point, and the set demonstrates variety in presuppositions ranging from modernist deductive logic (e.g., Fritz) to rhetorical ethical situatedness (e.g., Prellwitz), the collective insights of the essays invite clarity to emerge in what Martin Buber (1966) called "the between"—the delicate area between different people and their diverse positions. The ontological reality of the between contributes to all communicative parties involved, providing a common center that avoids the dangers of two extremes: the collective or "borg-like" mentality portrayed in episodes of *Star Trek* where humans are transformed into machines responding out of a central mind, and that of rogue individuality—individualism—in which a person or organization lives with a hyper-emphasis on the self (or organization) at the expense of others

(Arnett and Arneson 1999, 131). In the context of this work, both positions entrap one in a narrow, rigid understanding of communication ethics and crisis communication. They deny the complexity and reach of crisis into the past, present, and future and offer little hope for broadening the relationship between ethics response and crisis situations.

As editors, we work within Aristotle's commitment and step knowingly into a conversation begun by others. Therefore, we look to work that has already been done for starting points and foundational coordinates from which to engage the path laid in this book. Our approach is deliberate in its invitation to diversity of theoretical and methodological standpoints.

Within the public sphere, the starting points for crisis communication have their roots in public relations. Crisis decision making shapes an organization and its publics while considering multiple goods, making questions of communication ethics of paramount importance. Already, several volumes dealing specifically with crisis communication (e.g., Coombs 2007; Mitroff and Anagnos 2001; Ulmer, Sellnow, and Seeger 2007) address issues of ethics either directly or indirectly. Indeed, any discipline that deals with publics must treat the issue of ethics. For example, indirect treatments of ethics span both the public and private sphere with subjects such as truth telling, reputation, inclusion of multiple voices, and public accountability (Mitroff and Anagnos 2007; Coombs 2007). Fitzpatrick and Bronstein (2006) provide a more direct engagement of communication ethics in public relations but do not deal specifically with crisis communication and its ethical dimensions. Direct treatment of ethics occurs in Ulmer, Sellnow, and Seeger (2007) as they discuss managing communication ethically during a crisis and responding to the ethical demands of crisis.

Although the connection of a public relations understanding of crisis communication to communication ethics is important, we have already stated that the potential scope of the connection of crisis communication to communication ethics is quite broad. A cursory examination of books available in multiple domains of communication reveals some treatment of this connection, but it remains implicit for the most part. One book on interpersonal communication (McLean 2004) contains material on the "professional" dealing with crisis management in interpersonal interview settings rather than the nature of "interpersonal crisis communication" as an interpersonal phenomenon. Another treats interpersonal crisis (Donohue and Kolt 1992) as an element of conflict management, offering techniques for managing such crises. However, there is no treatment of ethics, and there is little theoretical rationale for the approach and offered techniques/strategies. Cupach and Canary (2000) offer a rationale for studying conflict: to promote "peace on earth" (6), which defines an implicit ethics, but with no "crisis" framing. To date, there is no volume on the market bringing together a collection of essays treating communication ethics and crisis communication from a broad range of topics.

The litmus test reveals an opportunity for this collection of essays amidst the nature of our postmodern world, replete with conflict over narratives, the good life and goods shaping that life, national differences, energy crises, scarcity of resources, and general uncertainty that prompt heightened awareness of difference and potential conflict. As Alasdair MacIntyre (1984) made clear in his classic treatise on the condition of our day, in a time of narrative contention, intelligibility must come from a holistic evaluation of a life (205). This is true for organizations as well as for persons and the spectrum in between, calling for an examination of crisis communication and communication ethics from the most intimate personal level (e.g., within the self) through the interpersonal and institutional (e.g., internal and external organizational communication), the political, social, and public. This broad panorama offers a heuristic view of this important and rapidly emerging topic. From this vantage point, we offer a treatment substantive in theory in a field often dominated by unreflective technique.

Our volume challenges the conversation to move beyond technique into the "between" of ideas about communication ethics, crisis communication, and their inevitable intersection—collision even—in everyday life. It does so through a thematic, story-laden approach grounded in theory hailing from multiple epistemological positions. The essays are grouped in Heideggerian terms (1996) according to their existential "usability" in the "between" of ethics and crisis. There are three sections that frame this resource for communication ethics and crisis communication: (1) "Narrativity" and "situatedness" (Benhabib 1992), (2) Response-ability as "inner transparency" (Ricoeur 1983, 89), and (3) the "association of memories" (Ricoeur 1983).

The firsts section deals with "narrativity" and "situatedness" and draws from an understanding of communication ethics and crisis communication as experienced within an immersive act "in a web of human relationships" (Benhabib 1992, 127). In this web, the self and other (crisis) are poised for understanding and engagement because of their situatedness. We are products of our experience and the narrative ground upon which we stand; this situatedness shapes our personal response to individual crisis that confronts humans by virtue of simply being in the world. Institutions must understand their own situatedness granted to them by virtue of their own mission. This standpoint provides a horizon out of which an appropriate and fitting response to crisis might emerge. Narrativity is the "mode" through which actions and response are understood and the nature of the crisis constituted for resolution and/or alkaline momentum (127).

- Mattson's essay engages crisis communication at the personal level through the communicative ethic of epideictic rhetoric. In his explication of Walker Percy's *The Moviegoer*, he explores situatedness as bringing forth a necessary response of being "at home" in the world yet not taking it for granted. For Mattson, the postmodern crisis of fragmentation is addressed through the dialectic of traditional verities and novel understandings of being in the world.
- Prellwitz urges consideration of Nietzsche's work—specifically the significance of "the death of god"—in order to unite dialogue and rhetoric in the service of today's communicative crisis of narrative and virtue contention. Through engaging what is ready to hand at any given moment of crisis we recognize the frailty and strength of our situatedness. Prellwitz argues that individual agency is always embedded within historicity, recognizing the situatedness of human action and its ethical implications.
- Seymour offers a phenomenological rendering of an intrapersonal "death of god" as she explores the reaches of anxiety as communicative crisis. Her essay delves into the "restless state of existence" as the universal human crisis. She offers an understanding of situatedness within language and the community of others to define an ethic of anxiety and offer coordinates for engaging the consummate restlessness that is the human condition.

The second section explores the idea of response-ability as "inner transparency" (Ricoeur 1983, 89). We recognize that crisis can never be comprehended or addressed fully. Ricoeur's (1976) understanding of appropriation and distanciation as elements of response suggests that one can never fully master crisis; there is always a productive distance between a responder and the crisis. Some element of ambiguity remains that cannot be overcome. We also believe that through the narrated world, a certain degree of inner transparency can be achieved. This form of transparency is borne out of a particular narrative. It works within temporality to provide a way of crossing time and generations vis-à-vis certain parallelisms that apply to given crisis situations. Inner transparency respects tradition while being fully engaged in the present historical moment. It anchors action within a broader history of actions and responds from the ethical juncture of understanding emergent from within that horizon. Inner transparency permits difference and stability to coexist, allowing for extraordinary flexibility within a narrative—one that must flex as historical moments change. These essays have a starting point resonant with a postpositivist and/or modernist deductive approach, moving to another point on the "narrow ridge" (Buber 1966, 110) to complement the initial set's dialogic, situated approach.

- Fritz addresses professional civility as an ethical response to interpersonal crisis in organizational settings. Her essay frames a theoretical and practical approach to professional civility as an interpersonal communicative ethic. Fritz establishes this ethic as a response emerging from the mission of a local organizational home, the story of professionalism as it has emerged historically in the United States, and respect for the nature of work itself.
- Groom "questions back" to the organization and its narrative as a necessary voice in crisis situations. The issue of response-ability is grounded reflectively in an organization's narrative commitments, which offer a framework for interpretation and forward movement. In times of crisis, Groom contends, an organization's narrative provides a hermeneutic entrance into the definition of the crisis itself, enabling productive distance and concomitant "biased clarity" necessary for ethical interpretation and action.
- The Sellnows and Venette take response-ability one step further by examining the nature of audience learning styles. They say that ethical communication of crisis messages involves crafting messages with attributes conducive to audience understanding and subsequent action. The Sellnows and Venette argue for inclusion of multiple learning style preferences in public messages related to crisis.

Finally, section three works within the idea of an "association of memories" (Ricoeur 1983), which refers to the importance of making visible the often unperceived distinctions between public and private contexts. Crisis in our current moment, in its broadest sense, is ubiquitous, confronting the human condition in multiple ways. It permeates public and private spheres to the extent that crisis has assumed a banality possible only through repeated encounters. Even though this banality of crisis is present, it heightens our awareness of the need for appropriate distinctions. What becomes public is not always what should be, and what has been hidden is often what should be known. Regaining time in a crisis situation is a matter of appropriate decision making and appropriate disclosure. We have forgotten a theoretical rationale for discerning what should be public and what should remain private. Crisis is not always a "problem" but an opportunity for creative response. These essays return to a rhetorical ethic based in situatedness, moving once more to the other side of the narrow ridge.

- Jovanovich and Wood explore how public declarations of justice take us to communication ethics. They evaluate Barack Obama's "a more perfect union" speech as a specific example of prophetic speaking, offering it as an example of public discourse that rallies a fragmented public sphere around common values.

- Gehrke looks at the opportunity presented by crisis for creative growth. His work proposes an understanding of crisis as the condition and constitution of living and communicating rather than as the exceptional moment foregrounded by most crisis communication theory and practice. Gehrke explores a constructive hermeneutic entrance into the interrelation of crisis communication and communication ethics, turning the unreflective into reflective engagement.
- Arnett concludes this volume with a comprehensive treatment of communication ethics and crisis management in a postmodern moment. The metaphor of "Janus at the gates" offers a situated understanding of postmodern communication ethics, encouraging an approach to crisis communication through a communicative ethic of learning from difference. Today's moment of narrative and virtue contention calls for public conversation about the role of private differences as we engage life together. Arnett challenges us to embrace this moment as a celebratory opportunity, moving forward towards an enlarged mentality in the twenty-first century.

And so our collection of essays extends the conversation on communication ethics and crisis communication to meet the challenges of a complex and ever-changing world. There is much to be gained from the multiplicity of perspectives presented here and from those yet to follow. We invite you to explore and continue the conversation.

REFERENCES

Aristotle. 1954. *Rhetoric*. Trans. W. Rhys Roberts. New York: The Modern Library.

Arnett, Ronald C. and Pat Arneson. 1999. *Dialogic Civility in a Cynical Age: Community, Hope and Interpersonal Relationships*. New York: SUNY.

Buber, Martin. 1966. *The Knowledge of Man: A Philosophy of the Interhuman*. New York: Harper and Row.

Coombs, Timothy. 2007. *Ongoing Crisis Communication: Planning, Managing, and Responding*. Thousand Oaks, CA: Sage.

Cupach, William. R., and Daniel J. Canary. 2000. *Competence in Interpersonal Conflict*. Prospect Heights, IL: Waveland Press.

Donohue, William A. and Robert Kolt. 1992. *Managing Interpersonal Conflict*. Newbury Park, CA: Sage.

Fitzpatrick, Kathy R. and Carolyn Bronstein. 2006. *Ethics in Public Relations: Responsible Advocacy*. Thousand Oaks, CA: Sage.

Heidegger, Martin. 1996. *Being and Time*. Trans. Joan Stambaugh. Albany: State University of New York Press. (Original work published 1953.)

MacIntyre, Alasdair. 1984. After Virute. 2nd ed. Notre Dane: University of Notre Dame Press.

McLean, Scott. 2004. *The Basics of Interpersonal Communication*. Boston: Allyn & Bacon.

Mitroff, Ian. I., and Gus Anagnos. 2001. *Managing Crises Before They Happen: What Every Executive and Manager Needs to Know about Crisis Management*. New York: AMACOM.

Ricoeur, Paul. 1976. *Interpretation Theory: Discourse and the Surplus of Meaning*. Fort Worth, TX: Texas Christian University Press.

——. 1984. *Time and Narrative*. Vol. 1. Trans. Kathleen McLaughlin and David Pellauer. Chicago: University of Chicago Press. (Original work published 1983.)

Ulmer, Robert Ray, Timothy L. Sellnow, and Matthew W. Seeger. 2007. *Effective Crisis Communication: Moving from Crisis to Opportunity*. Thousand Oaks, CA: Sage.

Part One

Narrativity and Situatedness

Impossible to Say

Walker Percy's Moviegoing Epideictic in Crisis Conditions

Craig E. Mattson
Trinity Christian College

Near the end of Walker Percy's 1960 novel, *The Moviegoer*, the book's most formidable character, Aunt Emily, interrogates her nephew, John Bickerson "Binx" Bolling, regarding his sexual liason with her stepdaughter.

> "Were you intimate with Kate?"
> "Intimate?"
> "Yes."
> "Not very."
> "I ask you again. Were you intimate with her?"
> "I suppose so. Though intimate is not quite the word."
> "You suppose so. Intimate is not quite the word. I wonder what is the word. . . ." (222)

The passage evokes a series of crises characteristic of modernity: crises of language, of morals, even of eroticism's diminishing returns. Aunt Emily's subsequent address may seem to some readers a gratifying form of crisis communication, eloquently responsive to predicaments both familial and broadly cultural, in which her "kind of folks have always possessed a native instinct for behavior, a natural piety or grace, I don't mind calling it" (222).

It is tempting to read this as an epideictic representation of what the book is really trying to say. But there are problems with this reading. The implied author has made it clear in the course of the novel that whatever else readers should distrust in Binx's first-person narration, he does have one sure ability: he can locate the devices of other people's inauthenticity. Because his characteristic response to bad faith in other parts of the novel has been a polite silence, his reticence here puts us on notice. In response to Aunt Emily's easy and authoritative speech, Binx's silence clears space for us to hear her

Americanization of Roman eloquence as smugly complacent discourse. We note her prolix familiarity with everything, her too-easy explanation of the world's anomalies. When Binx leaves her presence, neither repentant nor defiant but rather ridden with malaise, we are given to understand that her epideictic praises a dead way of life.

G. K. Chesterton once remarked that humans have a great need to be both at home in the world and astonished by it. It is a sad thing that Binx feels nowhere at home, but it is a far worse thing that nothing astonishes Aunt Emily. The astonishment at the heart of this novel is that Binx is able, somehow, to transcend both his aunt's spurious confidence and his own sick despair. It is difficult to name the farther stage he reaches—should we say he's become more centered, enlightened, whole, authentic?—without sounding like the moralistic professions of the radio show, *This I Believe*: "I believe in welcoming change, in giving lingering hugs, and taking long walks at sunset. This—I believe." Trying to fit Binx's experience into a narrative of private spiritual wholeness might make us vulnerable to the same lampooning that he gives the broadcast: "I believe in a good kick in the ass. This—I believe" (109). In fact, by the book's end, Binx himself seems unwilling to name where he is for the sake of others' edification, at least in part because he is uncertain whether "asskicking is properly distinguished from edification" (237). But what the novel's unreliable narrator is unable to say, except crudely, the novel's implied author achieves in full rhetorical measure. The implied author offers us readers not so much a narrative of private spiritual experience fit for inspirational broadcast as a species of epideictic that clears space for readers to share in the celebration of life's ampleness. This epideictic is not emergent from the utterances of the characters so much as from the rhetorical action of the novel's fiction.

EPIDEICTIC CRISIS IN POSTMODERNITY

Percy's creation of an epideictic novel involves the story in complex questions about the nature of ceremonial discourse in contemporary rhetorical theory. Epideictic discourse has always been difficult to define: even the ancients, those great categorizers, were unsure how to categorize this border-crossing discourse (Condit 1985, 284–285). In modern times, Kenneth Burke showed how epideictic blurs boundaries between aesthetics and rhetoric (1966, 295). Similarly, Chaim Perelman and L. Olbrechts-Tyteca noted that it smudges lines between education and propaganda (1969, 49, 51).[1] To deal with these complexities, Celeste Condit redescribed epideictic as a multi-functional, collaborative project in which speakers and audiences work on "communal definition"—a practical function in epideictic that is nonetheless irreducible to "the more narrowly argumentative and decision oriented functions served by legislative and judicial addresses, which are directed at more specific policies and goals" (1985, 291, 288). Narrowly argumentative or not, epideictic discourse, especially in Dale Sullivan's historical discussion, is still a richly purposive and social form of communication (1993, 116–117). He speaks of epideictic rhetoric's "ethos" in Michael Hyde's sense of the term as "the way discourse is used to transform space and time into 'dwelling places' (*ethos;* pl. *ethea*) where people can deliberate about and 'know together' (*con-scientia*) some matter of interest" (2004, xiii, xv–xvi).

The crisis of celebratory discourse centers on this question: is it possible, in a world of fragmented moral vocabularies and multifarious histories, to share epideictic space with others? (Ritivoi 2006, 103). Binx might respond, "It is impossible to say" (235). But this essay's reading of *The Moviegoer* suggests that the novel's implied author would respond with a tentative yes: *tentative*, because the old cultural consistencies by which Aunt Emily navigates the world have broken apart; *yes*, because this break-up creates new and hopeful opportunities for identity and solidarity for people like Binx.

EPIDEICTIC RECOVERY IN WALKER PERCY'S FICTION

At first glance, Percy seems an unlikely contributor to the conversations about epideictic. Unlike his counterpart in Southern Catholic letters, the neo-Thomist Flannery O'Connor, he was no Aristotelian.[2] It is not difficult to imagine his impatience, for example, with tidy divisions between deliberative, judicial, and epideictic genres, which he would probably label dismissively as *knowledge*—that is, information universally, categorically, and verifiably true. Instead, Percy professed interest in *news*, a messaging that speaks to the predicament of the alienated self (1975, 125–130). His clearly preferred literary companions in this quest for news—people like Kierkegaard, Marcel, and Sartre—were saboteurs of tidy conceptual frameworks. Their company, in fact, accounts for Percy's preoccupation with what seems a merely spiritual, merely private predicament: the malaise-ridden condition of the self, alienated, abstracted, displaced by modernity. Percy scatters these self-preoccupied souls across six novels—*The Moviegoer* (1961), *The Last Gentleman* (1966), *Love in the Ruins: The Adventures of a Bad Catholic at a Time Near the End of the World* (1971), *Lancelot* (1977), *The Second Coming* (1980), and *The Thanatos Syndrome* (1987). These protagonists seem too overwhelmed by a sense of the ineffable and the ironic in human experience to be capable of epideictic.

But one way to get at the rhetorical action in Percy's fiction is to examine the way his characters deal with vocational, familial, and psychological crises. The novel's favored term for the experience of these crises is *malaise*, a term also useful in the rhetorical tradition. Maurice Charland describes the condition as occurring when "a gap is felt between attributed social knowledge and personal experience, or indeed that particular conjunctures render social knowledge itself evidently contradictory or problematic"—a definition that returns us to the crisis of epideictic discourse in postmodernity discussed above (1999, 466). Percy's novel addresses this gap, although not in an immediately practical and decisional way. Instead, the novel's rhetoric is epideictic in Sullivan's sense of creating a place for consideration and for celebration.

Percy was not a rhetorician. He used the news/knowledge distinction not as a rhetorical category but as a philosophical (and sometimes theological) device for separating the existentialist sheep (who love news) from the behaviorist goats (who love knowledge). Even so, with a little jiggering, his distinction evokes a rough analogue in the rhetorical tradition, especially between those theorists who are inclined to trust epideictic discourse as participant in responsible social deliberation and those who are leery of epideictic's conservative ideological tendencies. The former group treats the materials that make up epideictic as a kind of knowledge—that is, as the unquestioned givens of community tradition, as the culture materials available for rhetorical action. The latter group treats the materials that make up epideictic address as ideological guy wires that trip up marginalized peoples before they ever reach the big tent of public discourse. Their critique of epideictic might be that it does not contain news in the sense of forwarding genuine social innovation for the good of the polis.[3]

These two approaches to epideictic action have temporal aspects that might go unnoticed apart from Percy's distinction between knowledge and news. Put in Burkean terms, Percy's distinctions clarify how these theorists' trust or distrust of epideictic embodies an attitude toward history. Those theorists who are alert to the epistemic capabilities of epideictic rhetoric—those interested in its knowledge—focus on the possibilities in the Now. They might ask, "Given current political and technological conditions, how can epideictic advance the good of the polis?" Contrarily, those theorists alert to the ideological tendencies of epideictic—those interested in news—tend to focus on the possibilities of the Not Yet. They might suggest that, given the injustice of current political and technological structures, epideictic action is likely to be spurious or exclusivist.[4]

But these rhetoricians' binary opposites become convertible forms in Percy's epideictic fiction. In *The Moviegoer*, Binx experiences the Now and the Not Yet simultaneously; he experiences a convertibility between the Familiar and the Strange. By positioning Binx "in a strange land where the signposts are enigmatic," Percy's epideictic fiction transforms at-homeness into astonishment and vice-versa (1975, 102). In hierarchical, homogeneous cultures, such a conversion would not have been necessary: epideictic address could preoccupy itself with the past and what it has to say about the present. But, as Ritivoi observes, "in a heterogeneous and pluralistic community that does not share an uncontroversial perspective on the past, epideictic must constantly justify its selection rather than count on being able to express existing values" (2006, 103). Binx's experience, shared with the reader, becomes just such a justified selection. Percy's approach to this crisis of epideictic is not to display an exemplary discourse: Aunt Emily's ceremonial speech, for example, is an eloquent failure. Nor does he retreat to the quirky private realm of a self-preoccupied narrator. Binx may be an oblique and untrustworthy communicator; but he is determined to stay with rhetorical action, that is, to "listen to people, see how they stick themselves into the world, hand them along a ways in their dark journey and be handed along, and for good and selfish reasons" (233).

Perhaps nowhere are these unstably good and selfish reasons more apparent than in Binx's aesthetic pursuits in the suburban moviehouse. To describe Binx's moviegoing, Percy borrows two terms from Kierkegaard, *rotation* and *repetition*; and these devices are part of a larger pursuit that he describes (in an idiosyncratic usage) as *certification*. These aesthetic devices serve as good reasons warranting Percy's epideictic discourse—not warranting in the sense of supporting praise or blame but rather in the sense of commending a way of comporting oneself with others in time. Rotation, repetition, and certification serve as "*elements that provide warrants for accepting or adhering to the advice fostered by*" the novel's epideictic (Fisher 1987, 48).[5] The advice this novel fosters is for the reader and the implied author to spend time together in a "timeless, consubstantial space carved out by their mutual contemplation of reality" (Sullivan, 1993,128). Through Binx's moviegoing devices, the novel's implied author skillfully negotiates a shared space with the book's readers where not only "a rediscovery of community and reconciliation," but also a kind of social ecstasy, becomes possible.

MOVIEGOING STRATEGIES AND EPIDEICTIC EXPERIENCE

Early in the book, Binx spies two uneasy newlyweds walking through the French Quarter. Just ahead, the movie star William Holden happens to appear, looking unsuccessfully for a light for his cigarette. When none of his admiring onlookers have a lighter, the newlywed man, with a studied casualness, proffers the light. It is an inspired action. Suddenly, he finds himself able to speak man-to-man with Holden and walks away transformed, having, as Binx notes, "won title to his own existence" (Percy 1961, 16). This is what Percy calls certification, or the charging of ordinary human experience with temporary significance, usually by the surplus importance of a celebrity or an expert. Percy's use of this device is not simply an account of an insular and quirky Binxian practice. It is an epideictic appeal to broadly felt experience, as recognizable as what happens in a studio audience when someone whoops at the mention of his or her hometown. The implied author's use of the device creates shared space with every reader for whom a mass-mediated shout-out has certified ordinary reality with importance.

For Percy, of course, the quest for certification is usually a register of alienation. His essay "The Loss of the Creature" in *The Message in the Bottle* argues that the self seeks certification when places and experiences have been emptied of significance by modernity. "Why is it almost impossible to gaze directly at the Grand Canyon . . . and see it for what it is—as one picks up a strange object from one's back yard and gazes directly at it?" (Percy 1975, 47). The short answer is that we have seen too many postcard epideictics, too many celebratory videos of the canyon. These mass-mediated images have created a "symbolic complex," a constellation of representations to which our direct experience must now accord. "[N]ow the sightseer measures his satisfaction *by the degree to which the canyon conforms to the preformed complex*" (Percy 1975, 47). Certification is thus a perversion of the shared cultural values usually considered essential to epideictic address.

Contributing to this same alienation is the culture of expertise. The displaced self, unable to encounter reality, settles for the approved constructions of technocrats: "The whole horizon of being is staked out by "them," the experts. The highest satisfaction of the sightseer (not merely the tourist but any layman seer of sights) is that his sight should be certified as genuine" (Percy 1975, 54). Put differently, the alienated self sees the world not for what it is but for how it is labeled. For such a self, the world looks like a hardware store whose shelves have spilled onto the floor, through which the self minces, managing every now and again to find a pipe fitting or a hexagonal wing nut that corresponds to a labeled plastic container on a shelf above her head. She has no idea what she is holding in her hand; she only knows what box it belongs in. Her chief aims are assortment and reallocation. "The worst of this impoverishment is that there is no sense of impoverishment. The surrender of title is so complete that it never even occurs to one to reassert the title" (54). Again, Percy's account of certification sounds like a perversion of a quality usually deemed necessary to epideictic: cultural authority.

Although Percy criticizes alienating tendencies in the certificatory quest, he suggests some virtue in the quest as well, especially as it reveals important yearnings in the human person. If a detached person like Binx can achieve a certain level of self-awareness about his malaise, there can be a kind of truthfulness in seeking certification: it can be a groping for grace, a fumbling for redemption of everyday things. This question can enable one "to live, for a time at least, as a person who is Somewhere and not Anywhere" (Percy 1961, 63). Certification by cinematic depiction offers an epideictic generosity, spilling energy over into quotidian life, diminishing alienation, creating significance. Binx's postmodern search for an energy to renew life with significance begins in the movie house where the possibility first becomes apparent that one's life could be ceremonially touched, one's experience opened epideictically.

Dale Sullivan asks, "Does epideictic, which once functioned to uphold a monolithic culture, operate in a postmodern world?" (qtd. in Ritivoi 2006, 97). To this, Percy's rhetoric of fiction gives a qualified yes. On a given day, certification looks (especially to a rhetorical critic) to be a legitimate rhetorical strategy, whereby a rhetor (like Binx) achieves agency in ways that no expert could have predicted. On that same day, from another angle, Binxian moviegoing looks (especially to an ideological critic) like an entrapment of the consumerist self in symbolic complexes erected by globalized mass media. It is impossible to definitively locate which angle offers the correct critique.[6] What we can do is to reword Sullivan's question slightly and ask *how* epideictic emerges in a postmodern world. Percy's moviegoing strategies of rotation and repetition help to answer that question.

There was a time when Binx Bolling was dismissive of repetition and rotation as strategies for dealing with the messiness of temporal human experience. Holding a modernist and hierarchical view of culture, he escaped malaise by pursuing a "vertical search"—reading Great Books and pursuing scientific abstractions—"and only for diversion took walks around the neighborhood and saw an occasional movie" (Percy 1961, 69). But increasingly exhausted by the depersonalizing devices of modernity, Binx attempts to certify his personal experience through rotation and repetition. As his approach to culture becomes, in good postmodern fashion, non-hierarchical, it also becomes more social. "But now I have undertaken a different kind of search, a horizontal search. As a consequence, what takes place in my room is less important. What is important is what I shall find when I leave my room and wander in the neighborhood" (70).

What he is looking for is a kind of aliveness, a celebratory expansion of human experience, akin to what audiences have sometimes felt hearing epideictic address. This opening in lived experience gives one the sense of self and society at once: this is who I am, this is who we are, and this is what our togetherness means. But not only is this experience painfully temporary, Binx quickly discovers that the problem that such an epideictic soothes the present without addressing the problem of his multifarious "pasts." His Aunt Emily offers one account of his life, his cousin Kate another, his war buddy from the Korean conflict another, and the photographs of his dead father another one entirely.

One solution to this welter of histories is to ignore the past altogether, especially by means of a strategy Binx refers to, in Kierkegaardian language, as *rotation*, or "the quest for the new as the new, the reposing of all hope in what may lie around the bend" (Percy 1975, 86). A great deal of commercial advertising proffers rotation to consumers: fast-food joints promising a new kind of chicken sandwich; car salesmen promising the latest model; movie trailers boasting cutting edge special effects. Ideological critics could well point out that global capitalism is dependent on such infinite promises of the new (Kernan 2004, 210). But for Binx, the messages of popular culture are not merely to be gawked at; they are semiotic portals—or, to offer a different figure, they are signposts of the possibility for "the experience of the new beyond the expectation of the experiencing of the new" (Percy 1961, 144). In such moments, time achieves a fullness that feels sacramental to the moviegoer.

Rhetoricians should appreciate the ways that Binx's moviegoing strategy enacts a kind of *phronesis* in the middle of epideictic discourse. By seeking certification, the moviegoer seeks a kind of judgment not unlike that of the rhetorician who, in Andrews, Leff, and Terrill's understanding, offers not only to evaluation but also "illumination" of contemporary discourse (1998, 6). This species of rhetorical judgment does not necessarily draw on formal, logical, or analytic criteria. Instead Binx invites that infusion of significance which Frederick Buechner describes as "that catch of the breath, that beat and lifting of the heart near to or even accompanied by tears, which I believe is the deepest intuition of truth that we have" (1977, 97).

As important as the present is to Percy's moviegoer, he is not satisfied to ignore the past altogether. Humans are storied creatures, after all, and cannot evade their addiction to history. Unfortunately, what complicates the past, and indeed the present as well, is the gloss of a mass-mediated symbolic complex. Binx mocks the way that such a symbolic complex overlays and vitiates even intense human experience, as when on a "summer night one met a lonely girl in Central Park and achieved with her a sweet and natural relationship" (Percy 1961,7). Binx's deliberately postcard-prosaic adjectives and his ironic descriptors all suggest a representational lacquer, created by an advice columnists, self-help books, and Harlequin Romances, which keeps people from authentic human experience. "I too once met a girl in Central Park," Binx adds, "but it is not much to remember" (7).

But instead of relinquishing the possibility of epideictic, Percy's rhetoric of fiction thrusts the protagonist into the middle of the prosaic in what Percy refers to as "dialectical savorings of the familiar as the familiar." There, genuine experience is "recovered from familiarity by means of an exercise in familiarity" (Percy 1975, 48). For Binx, recoveries of the familiar—he calls them *repetitions*—often occur in theaters. "The fact is," he notes, "I am quite happy in a movie, even a bad movie" (Percy 1961, 7). These sometimes banal familiarities merge his experience of being a self with the mediascape. Indeed, the theaters themselves enable him to feel at home in the world. On one occasion, he revisits a moviehouse he had attended fourteen years before and experiences "a secret sense of wonder about the enduring" (Percy 1961, 80). His describes this experience as dramatizing one's own history so that the interim between Then and Now can be concentratedly experienced "without the usual adulteration of events that clog time like peanuts in brittle" (Percy 1961, 80). Repetition, in other words, has to do with the deliberate return to the already experienced in hopes of recovering the essence of the experience.

Here Percy might seem to have commonality with ideological critics who would surely complain that repetition (not to mention rotation) is an incorporation in ideologies long rampant in the American mass media. From this perspective, Binx's engagement with mass-mediated *doxa*, understood as public opinion or as a set of cultural assumptions, is insufficiently alert to what McGee calls the "problems of *constraint*, investigating why, how, and with what result culture silences people" (1990, 71). Percy's implied author seems to share some of this discomfort with Binx's strategies: the moviegoer at the end of the story, after all, gets married and, after a fashion, gets religion—but without any mention of the movies that had once been so important to him. Percy himself seemed to have a deep ambivalence about moviegoing: "He is drawn to it always, as some of us are to McDonald burgers" (Montgomery 1993). Perhaps Percy, like his counterparts among the ideological critics, would say that certification offers little more than a cue to cultural fragmentation, boredom, amnesia.

But the moviegoer's aesthetic quests offer signposts indicating more than postmodern malaise. *The Moviegoer* also invites readers to consider how Percy's notions of rotation and repetition evoke a dialectic between newness and familiarity that speaks to a yearning deep in the human. Chesterton put the problem this way:

> How can we contrive to be at once astonished at the world and yet at home in it? How can this queer cosmic town, with its many-legged citizens, with its monstrous

> and ancient lamps, how can this world give us at once the fascination of a strange town and the comfort and honour of being our own town? (2009, 4)

There are moments in *The Moviegoer* when rotation and repetition, when strangeness and familiarity, become interchangeable, even convertible in a way that affords comfort and honor to the human. Those are moments when the novel enacts epideictic discourse in a postmodern key.

Take, for instance, one quiet-spoken scene late in the novel when Binx watches, in his rearview mirror, a black man walking out of a church on Ash Wednesday. The scene is easy to miss, because Binx is talking with Kate about something else entirely. He at first concludes that the man is doing nothing but a churchly version of repetition, a self-conscious social performance intended to impress whoever might be watching with the absolute at-homeness of the participant in sacramental procedure (Percy 1961, 233). Binx thinks he smells bad faith. The churchgoer is after the same thing as the moviegoer: religious ritual offers certification by the crackly impress of a bit of ash. But something about the scene and the symbolic action surprises Binx with an unlooked-for expansiveness, a mystery glinting at the edges of the mundane. He continues to watch in his rearview mirror, pondering:

> It is impossible to say why he is here. Is it part and parcel of the complex business of coming up in the world? Or is it because he believes that God himself is present here at the corner of Elysian Fields and Bons Enfants? Or is he here for both reasons: through some dim dazzling trick of grace, coming for the one and receiving the other as God's importunate bonus? (Percy 1961, 235)

Binx has encountered an epideictic symbol that does not operate in quite the way either a rhetorical critic or an ideological critic might expect. That is to say, this certification, this fingerprint of ash on the forehead, is not identifiable as an achievement of personal agency, a smart use of the available rhetoric for the purposes of *phronesis*. Nor is it definitively an unfortunate incorporation by up-and-coming middle-class *doxa*, the businessman's relinquishment of sovereignty for the sake of coming up in a capitalist world. Instead Binx wonders whether some importunate bonus has rendered the correspondence between rhetorical choices and consequences happily inexact. And that experience of benevolent asymmetry may be the necessary condition for epideictic in contemporary life. In an age when no cultural authority can indisputably set things right, it may well be that only gratitude for importunate bonus enables shared celebration across the numerous disparities of human life.

CONCLUSION

Percy wrote in the time when the greatest imaginable crisis was the Bomb, just as in our time we fear weapons of mass destruction. Rhetoricians, rightly convinced that rhetoric offers an alternative to violence, argue that if we can reconstruct a peaceful *sensus communis*, if we can foster generous practical reason, we may find an answer to the question of how to keep things from blowing up. But for Percy, a more disturbing question was, "What happens if the bomb doesn't fall?"

The short answer, of course, is that we go to the movies in quest of certification. We seek the strange by means of rotation; we seek the familiar through repetition. Rhetoricians might answer that Percy's "what-happens-if" question is too hypothetical to be worrisome. Ideological critics might point out, with a great deal of warrant, that persistent non-discursive conditions make universal peace an unlikely event any time soon. This indefinite deferral of peaceful conditions has at least the virtue of giving job security to rhetoricians: for the foreseeable future, there will always be needs to create identification across cultural lines. But even should we manage to forestall terrorist action and foster healthy rhetorical cultures across the globe, Percy's question quietly suggests that our moviegoing habits will continue. Why? Because improving rhetorical practice and ideological critique would not eliminate human experience of boredom and everydayness, arbitrariness and contingency. Percy's eschatological understanding of discourse thus asks us how to navigate the familiar without boredom and to engage the strange without amnesia. Binx's aesthetic quests for rotation and repetition tutor him—and through him, us—to look for a species of certification that is not so much an immanent satisfaction as a signpost of the transcendent. Such a certification proceeds not by the kind of epideictic address that Aunt Emily seems primed to give, but rather by opening human experience epideictically: new hospitalities arrive, a new habitation is at hand, new ecstasies are possible.

Let me offer one final suggestion for what this epideictic might look like today half a hundred years after *The Moviegoer* was first published. If Percy were writing his novel today, these rotatory and repetitive strategies would still be useful for creating shared epideictic space with readers. But the fragmentation of our mediated experiences suggests that his book should be called not *The Moviegoer*, but *The Movietrailer*. Lisa Kernan's study of the rhetoric of movie trailers hints at the continuing epideictic possibilities in even these scattered, abrupt, and associative texts. She notes that trailers offer us both consistency and innovation in ways that can "serve to remind us that just as trailers are not movies, the cultural experience of space, time and nation under globalization is not the only reality" (Kernan 2004, 211). Movie trailers, like the movies they anticipate, create in us an eschatological yearning, an "anticipatory consciousness" that "is based on real hopes that like *outside* trailers, and indeed, outside cinema and other media texts as well" (211). This, or something like it, is what happened for Binx Bolling. By some dim, dazzling trick of grace, the human person's experiences of rotation and repetition cues a *logos* outside the cinematic text. They signpost a larger sacramental union of the rhetorical and the material in which the word made flesh makes generous habitation among us.

REFERENCES

Andrews, James Robertson, Michael C. Leff, and Robert Terrill, eds. 1998. *Reading Rhetorical Texts: An Introduction to Criticism*. Boston: Houghton Mifflin Harcourt.

Buechner, Frederick. 1977. *Telling the Truth: The Gospel as Tragedy, Comedy, and Fairy Tale*. New York: HarperCollins Publishers.

Burke, Kenneth. 1968. *Language as Symbolic Action*. Berkeley, CA: University of California Press, 1999.

Charland, Maurice. "Rehabilitating Rhetoric: Confronting Blindspots in Discourse and Social Theory." In *Contemporary rhetorical theory: A reader*. Eds. John Louis Lucaites, Celeste Michelle Condit, and Sally Caudill. 464–473. New York: Guilford.

Chesterson, G. K. 2009. *Orthodoxy*. Rockville, MD: Serenity Publishers.

Condit, Celeste Michelle. 1996. "Hegemony, Concordance, and Capitalism: Reply to Cloud." *Critical Studies in Mass Communication*, 13, no. 4 (December), web.ebscohost.com/ehost/pdf?vid =25&hid=9&sid=940ba0f9–0d00–4712-a175–29935f76e116%40SRCSM1 (accessed June 30, 2008).

——. 1985. "The Functions of Epideictic: The Boston Massacre Orations as Exemplar." *Communication Quarterly*, 33, no. 4 (Autumn), web.ebscohost.com/ehost/pdf?vid=30&hid=9&sid =940ba0f9–0d00–4712-a175–29935f76e116%40SRCSM1 (accessed June 30, 2008).

Farrell, Thomas. 1991. "Practicing the Arts of Rhetoric: Tradition and Invention." *Philosophy & Rhetoric*, 24, no. 3, web.ebscohost.com/ehost/pdf?vid=28&hid=9&sid=940ba0f9–0d00–4712-a175–29935f76e116%40SRCSM1 (accessed June 30, 2008).

Fisher, Walter. 1987. *Human Communication as Narration: Toward a Philosophy of Reason, Value, and Action*. Columbia, SC: University of South Carolina Press.

Hyde, Michael J. 2004. Introduction to *The Ethos of Rhetoric*, ed. Michael J. Hyde, xiii–xxviii. *Studies in Rhetoric/Communication*, ed. Thomas W. Benson. Columbia, SC: University of South Carolina Press.

Kernan, Lisa. 2004. *Coming attractions: Reading American Movie Trailers*. Austin, TX: University of Texas Press.

Marty, Martin. 2003. "Our religio-secular world." *Daedalus* 132: 42–48.

McGee, Michael Calvin. 1990. "Text, Context, and the Fragmentation of Culture." *Western Journal of Speech Communication* 54, 3 (Summer), web.ebscohost.com/ehost/pdf?vid=3&hid=5&sid=673e1433–603e-41fe-84fe-5e2f0f60abcf%40sessionmgr4 (Accessed April 1, 2010).

Montgomery, Marion. 1993. "Walker Percy and the Christian Scandal." *First Things*, 32 (April), www.firstthings.com/article/2008/08/004-walker-percy-and-the-christian-scandal-20 (accessed April 1, 2010).

Percy, Walker. 1975. *The Message in the Bottle: How Queer Man Is, How Queer Language Is, and What One Has to Do with the Other*. New York: Noonday Press.

——. 1960. *The Moviegoer*. New York: Vintage.

——. 1991. *Signposts in a Strange Land*. Ed. Patrick Samway. New York: Picador.

Perelman, Chaim and L. Olbrechts-Tyteca. 1969. *The New Rhetoric: A Treatise on Argumentation*. Trans. John Wilkinson and Purcell Weaver. Notre Dame, IN: University of Notre Dame Press.

Ritivoi, Andreea Decciu. 2006. *Paul Ricoeur: Tradition and Innovation in Rhetorical Theory*. SUNY series, Rhetoric in the Modern Era, Series eds. Arthur E. Walzer and Edward Schiappa. Albany, NY: State University of New York Press.

Sullivan, Dale. 1999. "Identification and Dissociation in Rhetorical Expose: An Analysis of St. Irenaeus' A*gainst Heresies*," *Rhetoric Society Quarterly*, 29, no. 1 (Winter), web.ebscohost.com/ehost/pdf?vid=30&hid=9&sid =940ba0f9–0d00–4712-a175–29935f76e116%40SRCSM1 (accessed April 1, 2010).

——. 1993. "The Ethos of Epideictic Encounter." *Philosophy & Rhetoric*, 26, no. 2 (Spring), web.ebscohost.com/ehost/detail?vid=4&hid=5&sid=673e1433–603e-41fe-84fe-5e2f0f60abcf%40sessionmgr4&bdata=JnNpdGU9ZWhvc3QtbGl2ZQ%3d%3d#db=ufh&AN=16242453 (accessed April 1, 2010).

NOTES

1. But the Belgians may have been too preoccupied with critiquing Enlightenment dismissals of rhetoric to offer a sufficiently rich postmodern account of epideictic. Celeste Condit critiques them for tending "always to subsume them [epideictic discourses] under judicial and legislative rhetoric" with the result that too much rhetorical agency was given to the speaker and not enough to the audience (1985, 286).

2. Something of Percy's relationship to the Greco-Roman tradition can be traced in his essay collection *Signposts in a Strange Land*, especially in his essay "Stoicism in the South."

3. This reading of the rhetorical tradition is an adaptation of Charland's examination of the aims of rhetoricians and critical theorists in "Rehabilitating Rhetoric." Few rhetoricians will tuck neatly into either of these two camps, one earnestly professing

epideictic's competency to use knowledge to foster civic wisdom and wellbeing, the other stoutly preaching the news of epideictic's ideological limitations.

4. This eschatological description of the rhetorical tradition and its relationship to ideological critique is informed by a reading of exchanges in *Critical Studies in Mass Communication* between Celeste Condit & Dana Cloud regarding Condit's "theory of concordance." See especially "Hegemony, Concordance, and Capitalism: Reply To Cloud."

5. For more on the need for coping devices in a diffusely mediated society see Gitlin, Todd. *Media Unlimited: How the Torrent of Images and Sounds Overwhelms Our Lives.* New York: Henry Holt and Company, 2002; Abercrombie, Nicholas and Brian Longhurst. *Audiences: A Sociological Theory of Performance and Imagination.* London: Sage Publications Ltd, 1998; Zengotita, Thomas de. *Mediated: How the Media Shapes Your World and the Way You Live in It.* New York: Bloomsbury, 2005.

6. I derive inspiration for this idea of the convertibility of the rhetorical and the material from Martin Marty's exploration of the convertibility of the sacred and the secular in his article "Our Religio-Secular World": "Individuals, cultures, nations, and societies, in certain contexts, dimensions, and perspectives, are really secular, in any plausible definition of the term. At the same time they are, in another dimension and from another perspective, really religious, in almost unguarded ways" (2003. 47).

Dialogic Meeting of Crisis

Illuminating Illusions of the Death of God and Nietzsche's Revaluation of Values

John H. Prellwitz
University of Pittsburgh at Greensburg

The present essay articulates a communication ethic to engage difference in a constructive fashion. Difference is positioned as a major influence that drives contemporary narrative and virtue contention in an age that Alasdair MacIntyre persuasively designates "after virtue." The ground for this constructive communication ethic arises from interpreting otherwise one of the subterranean springs of our contemporary postmodern historical moment, Friedrich Nietzsche. Dialogue serves as the alternative hermeneutic entrance taken into Nietzsche's work rather than engaging his writings from the standard deconstructive bias. Dialogue, it will be argued, not the monologic, mechanistic unmasking of indeterminacy within texts of human language, animates the architectonic rhetoric of his communication ethic. Dialogue and rhetoric, approached through their primary interpenetrating function, the disclosure and articulation of meaning, emphasizes the primary role of this constructive communication ethic making meaning communicable through whatever forms the temporal, existential situation avails, through word, gesture, or act.

Approached from this dialogic perspective, the meeting of difference embedded in competing values, morals, and traditions affords opportunities for questioning, interpretation, criticism, judgment, and action. This perspective opens one to the recurring interplay of dialogue and rhetoric that simultaneously gives meaning to our words and deeds. Opportunities discovered through interpreting otherwise Nietzsche's prophetic provocations on the death of god and the impending nihilism resulting from this penultimate loss of meaning and value then reveals the architectonic rhetoric of a constructive communication ethic, dialogic meeting. Dialogic meeting, the articulation of interrelationships of Nietzsche's guiding metaphors of perspectivism, genealogy, and revaluation of values, describes Nietzsche's communication ethic.

Nietzsche's ethic emerges as a response to the calls of crisis meeting the ills of his age through a renewed dialogue with, not only a deconstruction of, tradition, in the contemporary and eternally recurring contest of competing values. Positioned at the turning from modern to postmodern, Nietzsche even serves as a key protagonist to Alasdair MacIntyre's contemporary call to respond to the crises in ethics of our own historical moment of narrative and virtue contention. In contrast to Nietzsche, MacIntyre offers a response grounded in a narrative of the perennial importance of Aristotelian tradition and noble virtues in morality and ethics in his influential works *After Virtue* and *Three Rival Versions of Moral Enquiry* and continuing through his latest work, *Dependent Rational Animals*.

Also, MacIntyre's works disclose a consequence of the crises with which Nietzsche struggled that continues to influence the contemporary postmodern landscape within the discourse on morality and values, the loss of universals and traditions to guide judgment (Lucaites and Condit 1999). Stuart Sim states that the leading contemporary philosophical style "is 'antifoundational'" wherein one disputes "the validity of the foundations of discourse, asking such questions as 'What guarantees the truth of your foundation (that is, starting point) in its turn?'" (2001, 3). The struggle for a substantive standpoint from which to enter the conversation on values becomes paramount. How do we interpret, articulate, and enact a communication ethic in a world shaken to its cornerstones by emotivism, deconstruction, and the traumatic terror and wars of the past two centuries recurs throughout the shifting currents of the postmodern conversation. Sim continues,

> Postmodernism has drawn heavily on the example set by antifoundationalist philosophers, perhaps most notably the iconoclastic nineteenth-century German philosopher Friedrich Nietzsche, whose call for a 'revaluation of all values' constitutes something of a battle-cry for the movement. (2001, 3)

Nietzsche, often hailed as a precursor of poststructuralism and postmodernism, therefore stands as a foundation for antifoundational contemporary thought.

However, the problem of articulating a constructive communication ethic within the contemporary postmodern condition is complex. The topic rises to the fore of MacIntyre's influential study, *After Virtue*. MacIntyre therein states,

> What we possess . . . are the fragments of a conceptual scheme, parts which now lack those contexts from which their significance derived. We possess indeed simulacra of morality, we continue to use many of the key expressions. But we have—very largely, if not entirely—lost our comprehension, both theoretical and practical, of morality. (1984, 2)

Regarding the practical enactment of moral ideals and traditions through their communication and aesthetic, architectonic rhetorical constructions, MacIntyre later reports that "there seems to be no rational way of securing moral agreement in our culture" (1984, 6). If no foundations remain that gain communal consent upon the meaning of good and evil, truth and lie, how could agreement on guiding values for thought, speech, and action be reached? The difficulty of communal consent amid political, corporate, religious scandals of recent years and rising tides of unreflective cynicism and emotivism, claiming that "all moral judgments are *nothing but* expressions of preference," only reinforces the severity of the problem (MacIntyre 1984, 12). If correct, then the undecidability of such fundamental human questions of truth or lie, right or wrong or the ground upon which to decide such questions to guide word and deed would designate postmodernity as after morality, as after virtue in at least two important ways. First, postmodernity exists temporally as an age after virtue, after the tradition of virtue ethics has lost its power to claim public adherence and its ability to inform personal morals. Also, postmodernity discloses the opportunity to seek after virtue, to engage the struggle for meaning and value, with the understanding that any tradition or narrative one consults to offer value must be embedded within the mud and mysteries of the existential moment of decision—a moment that such recognition demands we meet anew, always open to the discovery and meeting of difference as an eternally recurring opportunity to create meaning from meeting crisis as difficulty and as a call to decision and subsequent communicative activity.

One way out of this conundrum is to engage in further investigation of dominant influences upon our current legitimation crisis. Jean-Francois Lyotard opens a fruitful path of inquiry by reminding us of the textured present in which we eternally reside, our contingent human condition. Lyotard posits that "*postmodern* would have to be understood according to the paradox of the future *(post)* anterior *(modo)*" (1999, 81). Therefore, Lyotard characterizes the postmodern condition as bereft of the traditional metanarratives that previously justified values and ethical reasoning. This understanding characterizes the postmodern present as simultaneously always adrift amidst the currents of the past and the unspoken possibilities of the future.

Situating itself within this tension, the present essay takes one route to address the problems posed by our contemporary context and its associated crisis of values and morals. Investigations open upon a major source of suspicion that plagues judgment between competing value claims today, the maelstrom that howls in the wake of Nietzsche's provocations on the death of god and his authored response through his revolutionary revaluations.

Nietzsche's notion of moral inquiry, genealogy, evolves into revaluation as an eternally recurring invitational dialogue demanding that individual agency be perceived as a contextual element always embedded within historicity. Drawing from the diversity of narratives calling for the opportunity to serve as the foundation for future action, Nietzsche authors a drama of ethics and moral reasoning through the play of epideictic rhetorical prose. "His brilliant style does not hide or cloud issues but dramatizes them" (Solomon 1988, 8). The present study seeks to explore and explicate the play and turning points of the plot of ethical argument in our lives from the vantage of Nietzsche's thought embedded within the storm and stress of authoring an ethical life.

The present essay begins by undertaking a genealogy of the idea of a dying god in regard to Nietzsche's statements. Although perhaps most often linked with the idea of the death of god in the contemporary public imagination, there are potentially many springs from Nietzsche's philosophical and religious background as the son of a Lutheran pastor and from his philological and philosophical training and tradition that could influence his provocation. Classical sources range from Nietzsche's readings of old Germanic, Scandinavian, and Greek mythology and the writings of Plutarch. For example, Heidegger states that the strange notion of the death of a god and the dying of the gods was already familiar to the younger Nietzsche. In a note from the time of the completion of his first work, *The Birth of Tragedy*, Nietzsche writes (1870): "I believe in the ancient German saying: 'All gods must die.'" (Heidegger 1977, 58) Blaise Pascal, of whom Nietzsche himself confessed "the fact that I do not read but *love* Pascal, as the most instructive victim of Christianity" (1967, 3), provides another link to the idea of a dying god in the Hellenistic world with his quotation of Plutarch's epitaph, "*Le grand Pan est mort* ['Great Pan is dead'] (*Pensées*, 694)" (Heidegger 1977, 59). However, perhaps Philip Mainländer's *Philosophie der Erlösung*, written in 1876 and read by Nietzsche in 1876–77 and 1883, may serve as a signal influence (Brobjer 2008, 89). Mainländer proclaims that "God has died, and his death was the life of the world" (as qtd. in Brobjer 2008, 89). However, Mainländer appears only once in Nietzsche's published corpus and according to a note on this passage in *The Gay Science*, Kaufmann states that Nietzsche cites him only to show the lack of influence of Schopenhauer's pessimism on German philosophy (1974, 309).

Many other pronouncements of the demise of god also surfaced in nineteenth-century German intellectual culture. Max Stirner writes that "man has killed God in order to become now—'*sole* God on high'" in his 1844 work, *The Ego and His Own* (139). Additionally, Heinrich Heine, a poet Nietzsche greatly admired, announced in his ironic 1835 piece *On the History of Religion and Philosophy in Germany*, "Do you hear the bell ringing? Kneel down—They are bringing the sacraments to a dying God" (2006, 266). Von Der Luft cites

Immanuel Kant as the one at whose foot Heine lays blame for killing god. He states that Heine described Kant as "this great destroyer in the realm of thought" and whose critical philosophy served as "the sword with which deism was slain in Germany" (Von Der Luft 1984, 269). With the claim that "God is noumenal," Kant destroyed the logical grounding of the three major proofs for God's existence, and we, "for whom all knowledge arises from experience, can know nothing of God's existence or nature," making faith in such a deity "absurd" (1984, 269). Benson also cites Kant's rationalistic reconfiguration of Christianity in *Religion within the Boundaries of Mere Reason* (1794) as a possible influence on Nietzsche's ideas on a dying god (2008, 30).

Benson continues the genealogical inquiry with other potential springs of influence upon Nietzsche such as:

> David Strauss's *Life of Jesus* (1835), in which he argued for mythical interpretations of Jesus' life and miracles and then, in *The Old Faith and the New* (1872), proclaims the "new" faith to be simply Christian moral teachings. Or mention Ludwig von Feuerbach's *The Essence of Christianity* (1841), in which the concept of God is reduced to a composite of human characteristics. But there was a whole host of German philosophers, theologians, and biblical exegetes who were putting both Christian theology and biblical writings under close and critical scrutiny. (2008, 30)

Heidegger also comments upon the presence of the death of god in Hegel's thought as yet another potential influence on Nietzsche's thinking (1977, 59). Von Der Luft states that the phrase "God is dead" occurs at least three times in Hegel's writings (1984, 263). In agreement with Heidegger, Von Der Luft notes the occurrence in Hegel's early treatise *Faith and Knowledge* (1802):

> The pure concept, however, or infinity, as the abyss of nothingness in which all being sinks, must characterize the infinite pain, which previously was only in culture historically and as the feeling on which rests modern religion, the feeling that God Himself is dead, (the feeling which uttered by Pascal, though only empirically, in his saying: Nature is such that it marks everywhere, both in and outside of man, a lost God), purely as a phase, but also as no more than just a phase, of the highest idea. (1984, 2: 266).

The phrase "God is dead" recurs also later in Hegel's *Phenomenology of the Spirit*. In the section titled "Revealed Religion" Hegel states that

> It is the consciousness of the loss of all essential being in this certainty of itself, and of the loss even of this knowledge about itself—the loss of substance as well as of the Self, it is the grief which expresses itself in the hard saying that "God is dead." (1977, 455)

These sources provide an overview of potential influences on Nietzsche, especially within his German cultural milieu. One additional source must now be examined before moving onto explication of Nietzsche's own statements on the death of god: the words of Ralph Waldo Emerson.

Walter Kaufmann states in his translator's introduction to *The Gay Science* that "Emerson was one of Nietzsche's great loves ever since he read him as a schoolboy" (1974b, 7). Nietzsche cited Emerson as only one of four nineteenth-century writers, the only non-European, "worthy of being called masters of prose" (1974, 92). Nietzsche also published the first edition of *The Gay Science* with an epigraph that paraphrased a line from Emerson's essay "History," and the title of Nietzsche's work itself may be inspired by Emerson's self-description as "a professor of the Joyous Science" (Kaufmann 1974b, 8). Higgins concurs with this genealogical rendering of influence (2000, 100). When discussing Emerson's possible influence on Nietzsche's thinking on the death of god, she states, "Having cherished Emerson since childhood, Nietzsche had surely read Emerson's lament [taken from his "Divinity School Address"] "Men have come to speak of the revelation as somewhat long ago given and done, as if God were dead" (Higgins 2000, 100). Collectively, these sources provide a glimpse into elements of the culture and climate of religion out of which sprang Nietzsche's diagnosis. It is to Nietzsche's writings on the death of god that we now turn to engage this penultimate crisis of life, meaning, and value within human existence at the rupture between modern and postmodern historical moments.

Nietzsche first articulates his meditation on the state of the Christian God within his historical moment in his work, *The Gay Science*. While Nietzsche first explicitly announces the death of god in section 108 of this work, it is his much more developed presentation in section 125 with which we will begin our exegesis.

> *The madman.*—Have you not heard of that madman who lit a lantern in the bright morning hours, ran to the market place and cried incessantly: "I seek God! I seek God!"—As many of those who did not believe in God were standing around just then, he provoked much laughter. Has he got lost? asked one. Did he lose his way like a child? asked another. Or is he hiding? Is he afraid of us? Has he gone on a voyage? emigrated?—Thus they yelled and laughed.
>
> The madman jumped into their midst and pierced them with his eyes. "Whither is God?" he cried; "I will tell you. *We have killed him*—you and I. All of us are his murderers. But how did we do this? How could we drink up the sea? Who gave us the sponge to wipe away the entire horizon? What were we doing when we unchained this earth from its sun? Whither is it moving now? Whither are we moving? Away from

> all suns? Are we not plunging continually? Backward, sideward, forward, in all directions? Is there still any up or down? Are we not straying as through an infinite nothing? Do we not feel the breath of empty space? Has it not become colder? Is not night continually closing in on us? Do we not need to light lanterns in the morning? Do we hear nothing as yet of the noise of the gravediggers who are burying God? Do we smell nothing as yet of the divine decomposition? Gods, too, decompose. God is dead. God remains dead. And we have killed him."
>
> "How shall we comfort ourselves, the murderers of all murderers? What was holiest and mightiest of all that the world has yet owned has bled to death under our knives: who will wipe this blood off us? What water is there for us to clean ourselves? What festivals of atonement, what sacred games shall we have to invent? Is not the greatness of this deed too great for us? Must we ourselves not become gods simply to appear worthy of it? There has never been a greater deed; and whoever is born after us—for the sake of this deed he will belong to a higher history than all history hitherto."
>
> Here the madman fell silent and looked again at his listeners; and they, too, were silent and stared at him in astonishment. At last he threw his lantern on the ground, and it broke into pieces and went out. "I have come too early," he said then; "my time is not yet. This tremendous event is still on its way, still wandering; it has not yet reached the ears of men. Lightning and thunder require time; the light of the stars requires time; deeds, though done, still require time to be seen and heard. This deed is still more distant from them than the most distant stars—*and yet they have done it themselves*."
>
> It has been related further that on the same day the madman forced his way into several churches and there struck up his *requiem aeternam deo*. Led out and called to account, he is said always to have replied nothing but: "What after all are these churches now if they are not the tombs and sepulchers of God?" (Nietzsche 1974, 181–182)

This incredible tapestry of the rupture of metanarrative justification implies a manifold of meanings. First, Nietzsche places a radical responsibility upon all those who come upon this parable. However, while this act destroys, it also calls forth change. The murder of God demands a response. What may be done now that the justification, meaning, value of life has been destroyed in the decomposing of a murdered god? Nietzsche envisions the radical responsibility for this event as also the radical responsibility to honor its depth and significance, to carry the mantle of meaning through the ensuing shadows. Nietzsche demands a new vision, a revaluation of values, as an opportunity to create meaning within this defining crisis.

The revaluation of values is the metaphor whose activity most explicitly demonstrates a concern with rhetoric of values. Nietzsche utilizes epideictic rhetoric, drawn from the character of a speaker, following Aristotle's understanding of *ethos*, ethics derived from character, to highlight the rhetorical nature of the struggle between competing moralities.

Revaluation then emerges from a dialogic perspective on the genealogy of morals that it has undertaken to question and study. Revaluation serves as a component of perspectivism, which one must not deny because it is uncomfortable. The acknowledgement of the affective, unprovable, rhetorical foundations of truths is a requirement of intellectual honesty and rigor, one to be embraced as providing additional objectivity regarding the situated, contingent nature of truth in a given context.

For Nietzsche implies the necessity of a dialogue on revaluation as he states of perspectivism: "Because we seek knowledge, let us not be ungrateful to such resolute reversals of accustomed perspectives and valuations with which the spirit has . . . raged against itself for so long" (1967, 12). This acceptance of reversals discloses one's *pathos* of distance, one's perspective towards revaluation of values. A revaluation of values also calls forth a revaluation of reasons as a theoretical means of opening new vistas of investigation. Revaluation from such a perspectival bias allows one "to see differently . . . to *want* to see differently, is no small discipline and preparation of the intellect for its future 'objectivity'" (Nietzsche 1967, 12). The articulation of this perspective of revaluation allows for an unrestrained striving for the truth of a given question, founded upon the theoretical base of perspectivism.

> [T]he latter understood not as "contemplation without interest" (which is a nonsensical absurdity), but as the ability to *control* one's Pro and Con and to dispose of them illustrating and proving that one knows how to employ a *variety* of perspective and affective interpretations in the service of knowledge. (1967, 12)

Hence appears an opening for a more expansive view of argumentation as acknowledgement—acknowledgement of the biases of embedded perspectives seeking to persuade while simultaneously engaging in the dialogic struggle for power through revaluation.

Nietzsche, through the dialogic perspectival investigations of genealogy, clears the ground for epideictic rhetoric of revaluation. This epideictic rhetoric of morals accepts the inherent rhetoric of interpretation, perspectives generated through genealogical inquiry as dialogic meeting. Revaluation acknowledges the necessity of utilizing more than merely the dialectic sparring of pro and con. Rather, rhetoric of revaluation of values would seek to engage in dialogue with a multiplicity of voices. Thereby, in the enduring struggle to be heard, dialogue acknowledges and is guided by the acceptance of the necessity of multiplicity to grasp the unity of an interpretation, judgment, decision, and activity to further knowledge, speech, and action in the service of life.

The objectivity gleaned through genealogical inquiry guided by perspectivism grounded in dialogic meeting allows for the eternally recurring contest of competing values to struggle for adherence within revaluation conducted in an era after illusion. Revaluation becomes rhetoric in service of a dialogic ethic as it seeks to utilize its objectivity as a means of persuasion. Revaluation seeks to praise and blame to accomplish the persuasion of its hearers. Revaluation remains tied to perspectivism in that it seeks not to refute standpoints via traditional logical argumentation. Revaluation rather engages competing wills to power in dialogic meeting of difference, simultaneously affirming and opposing competing views. Thereby revaluation explores a greater quantity of views to approach the truths that characterize the situation or question under analysis, enhancing the Nietzschean objectivity of these perspectives.

Revaluation illustrates not the completed task but the ongoing questioning of presuppositions of accepted, embedded traditional viewpoints. Acknowledgement of the "fact that man is the *as yet undetermined animal*" (Nietzsche 1966, 62) informs questioning of the values and morals that guide human word and deed. Revaluation embodies a controlled will to truth, one that does not cede questioning. Revaluation as a manifestation of the will to truth questions with awareness as to what social impact the revelation of some questions and answers may hold for people unaccustomed to such an honest embrace of truth's significance as ugliness and beauty. For example, revaluation may be said to seek to provide greater objectivity and thereby greater clarity as to the nature of truth and lie through a genealogical moral inquiry into the origin of these concepts and their influence on our present values and ethics. Through its saying of no and yes to the values and ethics of our past as background, we respond in the fleeting foreground of the temporal existential moment. Through our response to the challenge to act we engage in dialogic meeting of the eternally recurring contest of revaluation of the values in which we dwell.

Revaluation, then, operates within the dual nature of construction. Revaluation functions as a constructive rhetorical hermeneutic arising out of the necessity of destruction as demanded by the perspectival human condition in which we are called to meet difference in dialogue. Revaluation grounded in dialogic meeting allows for answerability of the other as a no and/or yes to the recurring epideictic contests of the values in which we dwell and out of which we think, feel, speak, and act.

To understand Nietzsche's metaphor of revaluation of values, explication of its language and form, its epideictic character, is necessary. The epideictic rhetoric of revaluation influences Nietzsche's perspective as it orders the arguments it makes in destroying and building perspectives from which to speak and act and linking the activity of revaluation with the dialogic ethic that directs its rhetorical activity.

Nietzsche's task of revaluation of values seeks to author the translation of the human back into nature. With acknowledgement of the inescapable situatedness of the human person, revaluation enacts meeting of multiplicity and difference in dialogue after illusion amid metanarrative disintegration. The perspectival nature of reality demands genealogical investigation of values and morals competing in the void created by unmasking the illusion of metanarrative explanations of reality and justifications for action. Following this deconstruction of the illusions of past metanarratives through genealogy, Nietzsche seeks a return to the future through the past. As a means of embedding humans within the temporal and existential moment, he engages in dialogic meeting of difference within the present historical moment. The dialogic nature of this embedding activity engages epideictic rhetoric as a call for response and mutual engagement. Epideictic rhetoric constructs values revalued through their saying in the activity of meeting difference and a multiplicity of perspectives in dialogue. To understand the task of revaluation, this work turns to the perspective Nietzsche adopts in enacting a perspectival revaluation of values guided by genealogical insights of meeting difference in the saying of the said, the presencing of the past values that influence human communication and activity.

Nietzsche considers values as explainable through their rhetoric. Values hold the power to create meaning and to persuade of the veracity of their meanings in the social struggle for power and value. "Power might thus be construed as the standard and measure of values" (Kaufmann 1974a, 196). Values serve as motives and evaluations inextricably linked in a grammar and vocabulary of human explanations of life and reason. Perspectives are grounded in the reasons for their calling forth, for their discovery prompted by the estimation of their value in constructively meeting the present situation. Perspectives emerge from the play of forces of competing to direct attention, word, and action. Genealogy undertakes to explain these perspectives in terms of their values, to draw back the mask of their rationalizations and perceive who and what interests function as their advocate. Genealogy of perspectives undertakes to discern value. Genealogical inquiry into the nature of values debates with contesting perspectives and in so doing destroys, constructs, and reconstructs values. Meeting perspectives in question enacts such a conception of dialogue amid the competing interests of other perspectives seeking power in the temporal existential moment of decision.

Explication of these values and the perspectives they support and explain, along with the necessity of choice and action that values urge, compels the conversation then to become a play between competing values.

> Let at least this much be admitted: there would be no life at all if not on the basis of perspective estimates and appearances . . . Indeed, what forces us at all to suppose that there is an essential opposition of "true" and"false"? Is it not sufficient to assume degrees of apparentness and, as it were, lighter and darker shadows and shades of appearance—different "values," to use the language of painters? (Nietzsche 1966, 34)

Values then perceived as the ground of perspectives and hence the ground of action function as the topic of critique implicitly and explicitly. The response articulates values. The response in the aesthetic fashioning of values that can compel adherence and demonstrate power in itself silently bestows value upon the process, persons involved, and conclusions reached. In the end, the ground of rhetoric becomes valuation and revaluation. Dialogic meeting of the exigency seeks to shape the genealogical excavations of its values and meanings in the temporal existential moment and frame subsequent perspectives rendered and interpreted from an ever-changing vantage.

These introductions convey why Nietzsche attached such import to questions of value, because they enable meaning, perspective, reflection, and action as they interpenetrate any conscious act of communication, cognition, and activity. Values and valuation permeate even the marrow of humanity itself. "Perhaps our word 'man' still expresses something of precisely *this* feeling of self-satisfaction: man designated himself as the creature that measures values, evaluates and measures, as the 'valuating animal as such'" (Nietzsche 1967, 8). How this process of valuation works Nietzsche elaborates in a section of his notes collected in *The Will to Power*. Therein he states, "The apparent world, i.e., a world viewed according to values; ordered, selected according to values . . . The perspective therefore decides the character of the 'appearance'!" (1968, 567) In this way, values become knowable only through their employment. Nietzsche further embeds the veil of values that animates our interpretations as he claims that "all experiences are moral experiences, even in the realm of sense perception" (1974, 114). The sense of these values and perspectives we inherit from language, science, religion, and other traditions whose narratives organize these values into meanings, recognizable interpretations of meeting the chaos of reality in the temporal existential moment. The illusion one must always recall is that we read them into our experiences, and each is open to rhetorical revaluation.

Values characterize the situation of valuation as well as determining which values attain power to direct human attention, word, and action and in what manner, to whom, and for what reasons. Likewise, values characterize the stimuli being met in discourse: "Our values are interpreted *into* things" (Nietzsche 1968, 590). What value resides in gold except for the value individuals, societies, and markets convey upon it? The metal itself does nothing; it is not employed except in costuming and consuming, yet it commands incredible economic value, value not in or of itself. The value of honesty or lying is shaped by the parameters of the situation, taking bread to feed a starving child versus preserving rights of ownership—the highest-ranking value compelling the appropriate action from the perspective of the person confronted with the choice and demand for judgment. Whether the act denotes necessity or illegality depends upon the perspective from which the ranking value is read into interpreting the situation. In this way, values adhere to the human necessity of judgment and action living in community.

Given the organic quality of values and valuation inherent in human life in community, why does Nietzsche call for a revaluation of all values as he meets his own historical moment in the dialogue his writings enact? Reasons emerge from Nietzsche's genealogical inquiries into the dominant value perspectives and centers of his day, which he perceived as authoring only paths to nihilism and destruction. This only further speaks to power inherent in values as illuminating rhetorics of power found in meeting, in saying and said of competing perspectives in an era of multiplicity and difference after metanarrative disintegration.

It is the power of the values of his era that Nietzsche perceived as needing to be revalued. To revalue the values of his age required the prescription of a return to naturalize man back into nature, to naturalize the modern values and lead toward life, not its denial, in waking death of nihilism. As Nietzsche declares, "man would rather will *nothingness* than *not* will" (1967, 28). To redeem activity of willing within dialogic meeting of difference, Nietzsche inaugurates his revaluation of values. Revaluation announces the communicative activity he denotes as his task (1968). Revaluation of values as Nietzsche's task comprises all of his writings, which participate as sayings of the said to prompt revaluation toward a new dawn of values and valuation.

Nietzsche's daybreak of new morals and values ironically seeks not creation as much as it does the brilliance of honesty and truth taken from the natural world surrounding and bearing the human condition. What Nietzsche then calls for is naturalization of values and of an end of the flights of values from embedded reality to the chimerical worlds beyond reality as a means merely to deny life due to one's personal decadence. Revaluation provides a perspective on the central tensions Nietzsche posits, the competing values in an era after metanarrative disintegration and the death of god.

> All naturalism in morality, that is all *healthy* morality, is dominated by an instinct of life—some commandment of life is fulfilled through a certain canon of "shall" and "shall not," some hindrance and hostile element on life's road is thereby removed. *Anti-natural* morality, that is virtually every morality that has hitherto been taught, reverenced and preached, turns on the contrary precisely *against* the instincts of life—it is a now secret, now loud and impudent *condemnation* of these instincts. By saying "God sees into the heart" it denies the deepest and highest desires of life and takes God for the *enemy of life*. . . . (Nietzsche 1968, 4)

This passage sheds light on Nietzsche's task of revaluation of values. The metanarrative disintegration, which he narrates and, as such, participates in, discloses the decadence that had come to cloud values in his modern era. The appeal to God was not one that sought to affirm creation and one's responsibility towards life but one that eased one's disgust at the elements of life not to one's personal liking. Rather than meeting difficulty as opportunity, modern manifestations and values as calls to God operated more to judge and castigate difference to become the same, to speak only from the perspective of the said (Nietzsche 1968, 5–6). From this nihilistic perspective, the activity of dialogue and meeting, removed from the uniqueness of the temporal existential moment, draws forth no meaning, only its lack. Life becomes sleepwalking and repetition, not making meaning and meeting, living as passive vegetation and not communicative, cognitive, and physical activity. This castration of the human spirit in the name of obedience and ritualized automatism came not to signify spirituality as much as to signal silence and death.

The decadent silencing of the multiplicity of meanings, which burst forth in the activity of meeting within the demands of the temporal existential moment, render life as living death. "For a condemnation of life by the living is after all no more than the symptom of a certain kind of life" (Nietzsche 1968, 5). Condemnation of this sort disembeds the critic from the situation he or she criticizes, unmasking the pretension to knowing the unknowable, speaking with the insight of all of life from the perspective of a single part. "One would have to be situated *outside* life, . . . to know it as thoroughly . . . as all who have experienced it, to be permitted to touch on the problem of the *value* of life at all" (Nietzsche 1968, 5). If one speaks from an inaccessible perspective, one speaks untruth and thereby falsifies all that one's speech praises and blames. The impossibility of speaking from beyond an embedded perspective of the temporal existential moment leads Nietzsche to conclude "sufficient reason for understanding that this problem is for us an inaccessible problem" (1968, 5). To accept the situated nature of the human condition breeds alternative paths to meeting difference in dialogue. "When we speak of values we do so under the inspiration and from the perspective of life: life itself evaluates through us *when* we establish values" (Nietzsche 1968, 5). As the valuating animal, humans take on the responsibility of living life through meeting the moments and exigencies they encounter from the resources of their embedded perspective, the only one that life affords and the only one required to construct a response. For even the anti-natural morality of decadence Nietzsche condemns affirms life, illustrating the flaws of the untruth and errors resultant from denying the perspectival conditions of human life as valuation.

Revaluation of values then inaugurates a rebuilding of the values in which we dwell. Let the human condition be the perspective in which one experiences and lives life and frame one's perspectival understanding, words, and deeds.

> Let your gift-giving love and your knowledge serve the meaning of the earth. . . . Lead back to the earth the virtue that flew away, as I do—back to the body, back to life, that it may give the earth a meaning, a human meaning. (Nietzsche 1954, 2)

The human meaning Nietzsche advocates is found in the epideictic rhetoric of dialogic meeting of difference in the temporal existential moment. Dialogic meeting builds meanings and values within the situated perspective human reality and existence afford as opportunity. The opportunity of dialogic meeting allows one to make meaning and value and act situated within the relations that constitute human consciousness and communication, to meet the relations and related nature of reality calling for a response. Working from within his historical moment, Nietzsche calls for a return to life through a revaluing of the dominant values that deny the difference and multiplicity of perspectives, values, and meanings that color each moment.

Revaluation of values, the giving of a human meaning to the earth through recognition of one's situated perspective on earth, demands meeting of decadence, the form of denying life by denying its perspectival, limited, embedded nature.

> Indeed, this is *my* insight: the teachers, leaders of humanity, theologians all of them, were also, all of them, decadents: *hence* the revaluation of all values into hostility to life, *hence* morality—*Definition of morality:* Morality—the idiosyncrasy of decadents, with the ulterior motive of revenging oneself against life—successfully. I attach value to this definition. (Nietzsche 1967, 7)

The denial of perspective communicates a denial of life and revenge against the limits of human life as situated living and limited knowledge. Therefore, the epideictic rhetoric of decadents authors morality as a perspective to persuade further denial and decadent living. Living as revenge as it works from an unreal sense of agency. This illusion compels belief that as one spoke, one spoke for all of humanity although his or her individual words were addressed to a specific situation. Communicative activity undertaken from this perspective denies and silences difference as a metaphor of the relations that situate and author one's life and its meanings in meeting.

In contrast, meeting as revaluation offers a situating of one's discourse within the multiplicity of difference and relations that provide meanings to one's life as a continuing dialogue. Dialogue unfolds through and with the ways in which one constructs meaning out of encounters that the temporal existential moment affords. Dialogue enacted from this perspective considers the conditions that foster dialogue as important in ascertaining its possible meanings and value.

> One will ask me why on earth I've been relating all these small things which are generally considered matters of complete indifference . . . Answer: these small things—nutrition, place, climate, recreation, the whole casuistry of selfishness—are inconceivably more important than everything one has taken to be important so far. Precisely here one must begin to *relearn*. (Nietzsche 1967, 10)

The knowledge of one's embedded nature allows for greater insight into meeting, greater awareness of the forces at play and the sources of influence upon perspectives and their perception within the activity of dialogic meeting of difference and multiplicity after illusion amid metanarrative disintegration.

Nietzsche's metaphor of revaluation begins with rhetoric, advances to rhetorical exegesis of nature of truth and lie, and advances to Greek tragedy as affirmation of life in contrast to Platonic denial. Then through Zarathustra, the revaluation of life's values—Dionysian/Zarathustran values as against the prevailing code of Christianity—rewrite nature into the text of life, *homo natura*. "When will we complete our de-deification of nature? When may we begin to "*naturalize*" humanity in terms of a pure, newly discovered, newly redeemed nature?" (Nietzsche 1974, 109) Now it is necessary to enact these revalued values as a way to meet difference in a postmodern context, a way to guide rhetoric as a communicative activity, a saying of the said as dialogic meeting. Now emerges a meeting that affords an architectonic rhetoric of *bildung*, to harmonize the relation of individual to community through shared culture: the values and language that speak and animate life between persons embedded within the historicity and contingency of the human condition.

A genealogy of the perspective of revaluation as a stance toward dialogic meeting of multiplicity and difference explicates what Nietzsche understands the materials of revaluation to be: the rhetorical power of values and their influence on human word and deed. The metonymical connection between the form of revaluation and its content, values, serves as the subject of the following section as this work continues its exegesis of values as elements or common places of the epideictic interplay of dialogue and rhetoric in the communicative activity of Nietzschean revaluation.

"Morals, Nietzsche meets, through epideictic saying of the said as critique to create new values. Values drawn from classical sources are revalued when meeting the demands of decision in another temporal existential moment through the perspective offered by these classical values. Nietzsche selected classical values to aid his task of naturalizing values and morals to help translate man back into nature. "It is the *Saying* that always opens up a passage from the Same to the Other, where there is yet nothing in common. A non-indifference of one toward the other!" (Levinas 1996, 6) This utterance foreshadows key elements of Nietzsche's epideictic revaluation of values. Nietzsche engages revaluation of values with *pathos* of non-indifference towards those to whom he seeks to meet in dialogue through his writings after the shattering of illusions and metanarrative disintegration.

Drawing from genealogical inquiries into the values and moral perspectives, Nietzsche engages in dialogic meeting through his task of an epideictic revaluation of values. The argumentation supporting the truthfulness of Nietzsche's perspective articulates an architectonic rhetoric as acknowledgement. Revaluation of values begins with acknowledgement of the "fact that man is the *as yet undetermined animal*" (1966, 62). Nietzsche's perspectival characterization and approach to reality prompts a revaluation of values and ideas of self and agency. Word and deed result from relations that influence meeting difference and multiplicity in the perspective of a self, not from a fixed self, but one fluid and constantly confronting consciously and unconsciously meanings and values struggling to gain adherence. In this way the person engaging in revaluation of values as Nietzsche suggests is called to praise and blame, to author distance and attain perspective. Nietzsche considers such a grounded perspective as necessary to provide distance and to enable the discernment of difference and embedded activity in response.

To assist in how to perceive a healthy, situated perspective from a decadent, life-denying perspective, Nietzsche authors a dramatic metaphor of the consequences of how one meets reality. The notion of the drama of meeting Nietzsche narrates through his metaphor of the eternal recurrence. The eternal recurrence provides an existential temporal moment in which the reader is called to enact a perspective from which to meet the difference inherent in living life in the event of meeting as being. For Nietzsche, being, if at all knowable, can be known only as ever becoming (Nietzsche 1974, 370; 1967, 3). The situation of the eternal recurrence also demands accounting of one's values and morals and enacts an epideictic accounting through one's response to the questions posed. The scene titled "the great weight" by Nietzsche follows, as its rendering is the only way to communicate its possible meanings and illustrate how it elaborates upon the tenets of revaluation.

> What, if some day or night a demon were to steal after you into your loneliest loneliness and say to you: "This life as you now live it and have lived it, you will have to live once more and innumerable times more; and there will be nothing new in it, but every pain and every joy and every thought and sigh and everything unutterably small or great in your life will have to return to you, all in the same succession and sequence—even this spider and this moonlight between the trees, and even this moment and myself. The eternal hourglass of existence is turned upside down again and again, and you with it, speck of dust!"
>
> Would you not throw yourself down and gnash your teeth and curse the demon who spoke thus? Or have you once experienced a tremendous moment when you would have answered him: "You are a god and never have I heard anything more Divine." If this thought gained possession of you, it would change you as you are or perhaps crush you. The question in each and every thing, "Do you desire this once more and innumerable times more?"would lie upon your actions as the greatest weight. Or how well disposed would you have to become to yourself and to life *to crave nothing more fervently* than this ultimate eternal confirmation and seal? (Nietzsche 1974, 341)

The struggle Nietzsche imagines through this scene, the ordeal of meeting the demon's provocation of the eternal recurrence, announces central elements of Nietzsche's metaphor of revaluation of values. The question requires an evaluation of the perspective from which one meets the difference and multiplicity one encounters after metanarrative disintegration. The illusion of agency and one's functioning as a self-determining and self-propelled wheel independent of alterity and the situatedness of one's reality are deconstructed in the question demanding a response. The question requires accounting of one's acknowledgement of one's values and ideas as constructed through the awareness of one's perspectival existence and knowledge. One knows through one's perspective, one's situation, values, and ethics. Meeting enacts the epideictic rhetoric of one's dialogic meeting. One enacts one's architectonic rhetoric of building meanings through the praising and blaming of values. These values are encountered through the words and deeds of one's life lived in conversation with the cultural and social relations that texture one's communication and consciousness as directed through the pressures of the need to communicate in speech and activity.

The challenge that the eternal recurrence presents is acknowledgement after meeting metanarrative disintegration. Meeting this challenge discloses a revaluation of agency and an interpreting otherwise of modern subjectivity, understanding selfhood as meeting, not as individual expression of preference. The person exists through dialogic meeting of difference in the temporal existential moment. One's pathos of distance informs genealogical investigation of possible perspectives and values from which to speak and act. The textured and embedded nature of communicative activity constructs the path from which emerges one's ethos, one's self born in meeting difference in a dialogic ethic.

> The ethical has to do with *ethos* in its originative sense of a cultural dwelling, a mode or manner of historical existence, a way of being in the world that exhibits a responsibility both to oneself and to others. It is this that defines the bearing of self as ethical subject, whose subjectivity is always that of an intersubjectivity. (Schrag 1997, 101)

Schrag's rendering of subjectivity knowable only as intersubjectivity points to the heart of a dialogic ethic that Nietzsche anticipates. As a prophet of a dialogic ethic, Nietzsche calls forth deconstruction situated within his modern historical moment to meet the excesses of perspectives that he determined sought to deny and destroy life—perspectives whose danger, Nietzsche claimed, resided in their displacement outside of time and hence outside of human meaning, value, and reality.

Nietzsche as prophet of a dialogic ethic authors an epideictic rhetoric of values to guide the activity of meeting. As he states in a work subtitled "Thoughts on the Prejudices of Morality," he praises virtues to guide one's genealogical investigations that enable the construction and destruction of perspectives affirming or denying of life from which to live. "*Honest* towards ourselves and whoever *else* is a friend to us; *brave* towards the enemy; *magnanimous* towards the defeated; *polite*—always: this is what the four cardinal virtues want us to be" (Nietzsche 1997, 556). This naming of guiding virtues communicates values to ground his dialogic ethic. These virtues are not open to choice but are compelled upon us from the perspective of the cardinal virtues. The values and ethics that guide life are found in living. Life as dialogic meeting of difference demonstrates awareness of one's perspective as a life situated amid the difference and multiplicity of the existential and temporal moment after metanarrative disintegration. Such a situated perspective after illusion compels honesty as one lives the mystery of an unfinished dialogue one is invited to engage. What will one make of one's time—what a glorious challenge Nietzsche claims awaits our future heroes of the spirit and artists of human living in dialogic meeting of the mysteries of the temporal moment.

Now given the interpretation of the death of god as a crisis of meaning which Nietzsche affords a constructive response through revaluation of values, we are awakened to a new light. "Let us beware of saying that death is opposed to life. The living is merely a type of what is dead, and a very rare type" (Nietzsche 1974, 109). Given the manifold perspectives Nietzsche reveals in the death of a decaying belief, we come through a revaluation of values of human life and existence perhaps to understand "'God' as the moment of culmination: existence and eternal deifying and un-deifying" (Nietzsche 1968, 712). In the eternally recurring demands of the temporal existential moments that constitute human living, we are called by the responsibility for others and the difference that animates our experience to meet crisis. Perhaps we may learn that we are called to meet the death of God as an opportunity to enrich our faith and beliefs. Perhaps we may learn that we are never to forget that we are called to assume continually the radical responsibility to question. Perhaps we may learn that we are called to question to sustain our values and justify the meanings we draw from meeting difference in dialogue, for "Unless one dies, one brings forth no fruit" (John 12:24).

REFERENCES

Benson, Bruce E. 2008. *Pious Nietzsche: Decadence and Dionysian Faith.* Bloomington, IN: University Press.

Brobjer, Thomas H. 2008. *Nietzsche's Philosophical Context: An Intellectual Biography*. Urbana, IL: University of Illinois Press. *International Nietzsche Studies* 13.

Hegel, Georg Wilhem Freidrich. 1977. *The Phenomenology of Spirit.* Trans. A.V. Miller. New York: Oxford University Press.

Heidegger, Martin. 1977. "The Word of Nietzsche: 'God is Dead.'' In Martin Heidegger, *The Question Concerning Technology and Other Essays.* Trans. William Lovitt. New York: Harper & Row.

Heine, Heinrich. 2006. "On the History of Religion and Philosophy in Germany." In Heinrich Heine, *The Harz Journey and Selected Prose.* Trans. and Ed. Ritchie Robertson. London: Penguin.

Higgins, Kathleen Marie. 2000. *Comic Relief: Nietzsche's Gay Science.* Cary, NC: Oxford University Press.

Kaufmann, Walter. 1974. *Nietzsche: Philosopher, Psychologist, Antichrist.* 4th ed. Princeton: Princeton University Press.

———. "Translator's Introduction." In Friedrich Nietzsche, *The Gay Science.* Trans. Walter Kaufmann. New York: Vintage.

Levinas, Emmanuel. 1996. *Proper Names.* Trans. Michael B. Smith. Stanford, CA: Stanford University Press.

Lucaites, John Louis, and Celeste Michelle Condit.1999. Introduction. *Contemporary Rhetorical Theory: A Reader.* Eds. John Louis Lucaites, Celeste Michelle Condit, and Sally Caudill. New York: The Guilford Press. 1–18.

Lyotard, Jean-François. 1999. *The Postmodern Condition: A Report on Knowledge*. Trans. Geoff Bennington and Brian Massumi. Minneapolis: University of Minnesota Press.

MacIntyre, Alasdair. 1984. *After Virtue*. 2nd ed. Notre Dame: University of Notre Dame Press.

Nietzsche, Friedrich. 1966. *Beyond Good and Evil: Prelude to a Philosophy of the Future*. Trans. Walter Kaufmann. New York: Vintage.

——. 1997. *Daybreak: Thoughts on the Prejudices of Morality*. Trans. R. J. Hollingdale. Eds. Clark and Brian Leiter. Cambridge: Cambridge University Press.

——. 1974. *The Gay Science*. Trans. Walter Kaufmann. New York: Vintage.

——. 1967. *On the Genealogy of Morals and Ecce Homo*. Trans. Walter Kaufmann and R. J. Hollingdale. Ed. Walter Kaufmann. New York: Vintage.

——. 1968. *The Will to Power*. Trans. Walter Kaufmann and R. J. Hollingdale. New York: Vintage.

——. 1954. *Thus Spoke Zarathustra: A Book for None and All*. Trans. Walter Kaufmann. New York: Penguin.

——. 1968. *Twilight of the Idols and The Antichrist*. Trans. R. J. Hollingdale. London: Penguin.

1997. Schrag, Calvin O. *The Self after Postmodernity*. New Haven, CT: Yale University Press.

2001. Sim, Stuart. "Postmodernism and Philosophy." In *The Routledge Companion to Postmodernism*. Ed. Stuart Sim. London: Routledge. 3–14.

1988. Solomon, Robert C. "Introduction." In *Reading Nietzsche*. Eds. Robert C. Solomon and Kathleen Higgins. Oxford: Oxford University Press. 3–12.

Stirner, Max. 1995. *The Ego and His Own*. Trans. Steven Tracy Byington. Ed. David Leopold. Cambridge: Cambridge University Press.

Von Der Luft, Eric. 1984. "Sources of Nietzsche's 'God Is Dead!' and Its Meaning for Heidegger." *Journal of the History of Ideas* 65, 2 (April–June): 263–276.

Understanding Anxiety

The Crisis of Ethical Choice

Celeste Grayson Seymour
Duquesne University

Understanding anxiety is largely a matter of one's phenomenological focus of attention. One can attend to anxiety as a "problem" that needs to be overcome or managed by behavioral techniques or, as Kierkegaard (1957) does, one can attend to anxiety as a necessity for human growth and existence, naming it humankind's best teacher. This work aligns itself with the latter attention to anxiety by articulating the necessity of understanding anxiety as an existential phenomenon or a state of existence of language-using selves in a world of others. This project asks: what is it about the nature of the human self—consistent across all historical moments—that constitutes a seemingly restless state of existence? This essay begins to explore this question by recognizing that we are creatures who are bound by language and who are bound by others. Accordingly, the author suggests that what could be experienced as a "crisis" of an anxious existence of a language-using self in a world of others can, in phenomenological terms, be turned toward a radical reflexivity of responsibility and choice. For being bound by language and by others means having boundaries, without which the self could not be determined. The metaphors of *intention, imitation, identification*, and *communion*, along with the philosophies of Søren Kierkegaard, Edmund Husserl, St. Augustine, Kenneth Burke, and Emmanuel Levinas, will help articulate the role that this phenomenological understanding of anxiety plays in driving the ethical choice of the self.

INTRODUCTION

At the commencement of this project, the author asks: "How does existential restlessness or what we colloquially call anxiety contribute to self-experience and what we understand as the shifting construct of human identity?" Most importantly for this project, however, the author asks "Does being in a state of anxiety constitute a crisis of the self?" The author remains attentive to issues of cultural and historical disparities in approaching these questions. However, this essay engages a more universal understanding of anxiety, at its basic level as a manifestation of our lives together as language-using selves in a world of others. This project explores the existential phenomenological origins of anxiety from a linguistic perspective. In a word, language creates the conditions of existential anxiety. As such, anxiety becomes a necessary component of our ethical systems as human selves attempt to navigate meaning within temporality.

Søren Kierkegaard's work on anxiety will provide the hermeneutic entrance into understanding this phenomenon. Kierkegaard (1980), in *The Concept of Anxiety*, links this state to human existence in temporality, naming it as the moment of possibility and freedom of the self. From Kierkegaard's work, this project explores the implications of human consciousness of life together as linguistic creatures. Edmund Husserl's (1970) phenomenological method opens up understanding of human consciousness to the metaphor of *intention*. We act always with intention guided by our phenomenological consciousness of the life world. Intention points to the act of *imitation*, or the movement toward an exemplar or form, as is understood in Augustine's three-fold subjectivity framed by knowing, willing, and loving (2002). Augustine's participatory self points to the willfulness of the subject as driven by the intention of discourse. With the introduction of discourse as vital to the development of the subjective self, the essay turns to Kenneth Burke's (1984) theory of *identification* and piety within *Permanence and Change*. Human symbol use creates systems of meaning that reflect what we believe to be identification with the sources of our being. Levinas's (1985) philosophy further expands upon this system of identification with an exploration of the discourse of being. Being speaks not of identification but of *communion*; ironically, a communion shaped by non-reciprocal responsibility and radical alterity. From these metaphors, we begin to understand the nature of anxiety as a necessary element of self-consciousness within temporal existence. This essay concludes, then, with a reflection on what an "ethic of anxiety" would look like from an existential phenomenological perspective.

LANGUAGE, COMMUNICATION, AND THE CONSTITUTION OF THE SUBJECT

Prior to the commencement of the discussion, however, it is important to clarify this project's framing of language and communication in its relation to the constitution of the subject. From an existential phenomenological perspective, the discussion of language is framed by the presupposition that language refers not only to the spoken and written words but also the communal embodiment of the Word in discourse and action. The Word constitutes the existence of the human being. In order to explore the relation of language and communication to the notion of the self, this essay turns briefly to Calvin Schrag (1997) in *The Self After Postmodernity*.

In *The Self After Postmodernity*, Schrag counters the modern notion of the autonomous self as well as the postmodern deconstructivist notion of the death of the subject. Martin Heidegger in *Being and Time* presupposes the notion of language as a "house of Being;" however, this notion of language is one of it being constituted by authenticity (Heidegger 1992, 217). Authenticity refers to Dasein's first-person relation to itself, or the immediate relation. Ontologically, this is the structure that Heidegger called "mineness" [*Jemeinigkeit*] (Heidegger 1962, 42). Schrag agrees that the first-person relation to the self is important, yet he recognizes that we must analyze this relation in terms of embedded and embodied communicative practices, not solely in terms of authenticity. Contrary to modernist presuppositions of the autonomy of the self from existence, a framework emergent from Cartesian dualism, Schrag argues that the self cannot be autonomous from the world of existence. In the same vein, Schrag answers the impulse of postmodern deconstruction to declare the dissolution of self-identity through "the death of the subject" (Foucault 1970, 387). We *can* frame an understanding of the subject after the postmodern turn to language. As he articulated in his earlier works, Schrag argues that this move calls for a *rhetorical* turn, or understanding subjectivity as constituted within an existential community of practices (1986).

In order to sketch a portrait of the self, Schrag envisages four manifestations constituting subjectivity: the self in discourse, the self in action, the self in community, and the self in transcendence. These four areas in turn describe the "who experiences" of "Who is speaking?," "Who is acting?," "Who is responding to other selves?," and "Who stands within and before transcendence?" (Schrag 1997, 4). In other words, the self is constituted by particular communicative practices embodied within communities oriented toward transcendental understanding of the self and others by way of discourse. This introduction of the notion of transcendence is not without a recognition of the historical and cultural embeddedness of the subject in existence. In his text, Schrag offers the illustration of an X and Y axis of existential subjectivity. He orients "the profiles of discourse, action, community, and transcendence as ingredients of a synchronics of self-understanding" on a vertical axis, with "the cultural-spheres of science, morality, art, and religion as the moving figures within a diachronics of historical constitution" on a horizontal axis (Schrag 1997, 5). With this spatial metaphor, Schrag points to the fact that the whole of language constitutes both the historical, dynamic nature of linguistics, (or the change of language from moment to moment) and the synchronicity of multiple speech acts that constitute the meaning of the self.

With this metaphor as a frame for understanding the nature of language and communication as constituting the self, the author of this project presupposes that the four profiles of discourse, action, community, and transcendence are manifestations of what can be understood as the metaphor of the Word. Consequently, each theorist presented in this project speaks to the interplay of Schrag's diachronic communicative practices in historical and cultural contexts as well as the synchronistic elements constituting linguistic consciousness. The Word illustrates our self as embedded in the dialectical tension of transcendental traces of our identity (historical, moral, and communal) in the midst of existential choices and acts. This tension is not without an occasional moment of restlessness, and perhaps even despair, as human beings negotiate the interspaces of otherness and identification. It is within this metaphor of the Word that this project turns to the particular phenomenon known as anxiety and the role it plays in guiding a self toward ethical action in existence.

ANXIETY AS POSSIBILITY FOR ETHICAL CHOICE

In 1844, Søren Kierkegaard wrote *The Concept of Anxiety*, in which he explores anxiety as a pervasive and basic characteristic of human existence in temporality. His phenomenological treatment of anxiety is based on an ontological view of human being as tension-filled. Several characteristics of human existence emerge from this work. First, Kierkegaard (1980) claims that humans are a synthesis of the mind and the body that is united by the third, or the spirit. Spirit is the closest state of human existence to the eternal, for it is within the human spirit that human being is most immediate to itself. Kierkegaard likens the human spirit to "a dreaming" that is at once welcoming and frightening; for as the spirit brings the body and the mind into one space, it places them both in a position of tension with one another (1980, 43). Human experience oscillates between the temporal and the eternal because we are beings derivative of Being.

Second, Kierkegaard (1980) explores the relationship between innocence and sin. He writes "in innocence, man is qualified in immediate unity with his natural condition" (Kierkegaard, 1980, 41). It is within the moment of innocence that one becomes most aware of oneself. The awareness of the self comes not from temporal knowledge but instead what Kierkegaard calls "ignorance:" "Innocence is ignorance" (1980, 43). It is ignorance about nothing, or, rather, awareness of no thing. Kierkegaard (1980) claims that in this stage of innocence, there is no knowledge of good and evil. There is only knowledge of possibility. In ignorance, there are no boundaries, and there are no distinctions. Just as in the story of Adam and Eve in the Garden, the knowledge of good and evil—linguistic distinctions of the temporal world—is what propels human beings out of innocence and into sinfulness. Kierkegaard writes "sin is precisely that transcendence, that *discrimen rerum* [crisis] in which sin enters into the single individual as the single individual" (1980, 50). The temporality of the world, experienced through the body and the mind of the subject, is the original state of sinfulness for humans as language-using selves.

Third, Kierkegaard (1980) characterizes humankind by its drive to act simultaneously as an individual and an entire race, an inherent contradiction of our life as social creatures. According to Kierkegaard (1980), a contradiction is an expression of a task; a task is a movement in history. The history of our race, then, is one of existing in sinfulness. Kierkegaard's understanding of sinfulness is different from that of a psychological rendering of a state of mind. He writes: ". . . sin is not a state. Its idea is that its concept is continually annulled. As a state (*de potentia* [according to possibility]), it is not, but *de actu* or *in actu* [according to actuality or in actuality] it is, again and again" (Kierkegaard 1980, 15). Sinfulness is never a possibility; it is an actuality. Sinfulness is temporality, or the actualization of our ethical decisions with the temporal. Within history, humans are faced with continual tensions within and threats to their being. These crises are characterized by moments of transcendence where one does not rise above, but instead "falls" back within, temporal existence (i.e., sinfulness). This is what Kierkegaard calls a "qualitative leap," (1980, 37, 41) or an act of intention within temporality. The "leap," a breaking forth into the temporal and the ethical, makes sin possible. Sometimes we leap in the wrong direction. Sometimes we do not. It is the possibility of the leap itself that brings anxiety.

Anxiety is a stage that is necessary before one makes the leap of faith; it is the stage where one shudders at one's freedom. It is that pregnant moment of possibility, of becoming, when one is faced with the crisis of temporality. Kierkegaard (1980) claims that anxiety occurs in every human life prior to the actualization of ethical choice. Human ignorance brings peace and repose, for the spirit brings comfort. However, due to the nature of our temporal being, we cannot stay within this state of grace. Anxiety emerges within ignorance. When there are no boundaries and only possibility, the temporality of the self cries out for distinction. Kierkegaard claims that the self becomes "dizzy" with its own possibility (1980, 92). In *The Concept of Dread*, Kierkegaard calls this possibility the "I can" (1957, 44). The moment one recognizes "I can," or the moment when anxiety emerges, is "the pivot upon which everything turns" (Kierkegaard, 1980, 43). Anxiety begets consciousness as the temporal self searches for ground.

The development of consciousness is a crucial moment in the emergence of the ethical self. In *The Concept of Dread*, Kierkegaard (1957) uses the story of Adam to illustrate this phenomenon. In a sense, the myth is re-enacted by every human being somewhere between the ages of one and three, for every individual must awaken from the state of innocence into which he or she is born into self-consciousness. Just as the story of Adam portrays, each child develops knowledge of good and evil first, which gives rises to conscious choice. With this knowledge, there comes a heightened sense both of the nature of possibility and of the responsibility that goes with it. The child then begins to move toward individuation tempered by the environment. In *The Meaning of Anxiety*, Rollo May claims that "individuation (becoming a self) is gained at the price of confronting anxiety inherent in taking a stand *against* as well as *with* one's environment" (1977, 39). The metaphor of anxiety for Kierkegaard, then, involves how a person can will to be amidst the continual dissolution from Being. According to May (1977), Kierkegaard's contention is that the freedom of selfhood depends upon an individual's capacity to confront anxiety and move ahead within history not *despite* it, but precisely *because of* it.

Anxiety, as a moment pregnant with the possibility of being, allows for a reflective sense of the tension and crisis of human existence. It is the phenomenological predecessor to intentionality. Kierkegaard claims that anxiety "approaches sin as closely as possible, as anxiously as possible, but without explaining sin" (1980, 92). In a complex and dialectical relationship, moments of anxiety and sin, of the awareness of no-thing and of some-thing, texture the human condition. Kierkegaard's theories create a foundation for understanding the phenomenological origins of the idea of anxiety. From Kierkegaard, this essay now turns to key metaphors that emerge from attempting to understand this human anxious existence. The metaphors that will shape the remainder of this essay, echoing the works of Kierkegaard, include: (1) intention, (2) imitation, (3) identification, and (4) communion.

Intention

The issue of intention becomes important in understanding an anxious state because the individual cannot "stand outside" the threat or objectify it. As Kierkegaard (1980) points out, anxiety emerges because there are no boundaries of the self but only possibility. An individual's subjectivity is called into question as the anxious state of possibility strikes at the perception of a person as distinct from the world of objects and others. The distinction between subject and object can break down; the boundaries of the self blur or fade. Therefore, anxiety is often accompanied by feelings of powerlessness, hopelessness, and decreased self-awareness—in a way, a crisis.

From this notion of the crisis of anxiety, this essay turns to the work of Edmund Husserl and the notions of consciousness and intentionality. In *The Crisis of the European Sciences*, Husserl's (1970) phenomenological method begins with the claim that phenomenology involves consciousness, but it is always consciousness *of* something. It is an intentional consciousness; the objects and others that we experience in this world are experienced, in reality, as one's intending or attending toward that thing. One cannot understand existence as the phenomena/noumena, or subject/object divide, as Kant did (Husserl 1970, 91–97). Husserl claims that we do indeed primarily experience the world through the physical body, or what he calls "kinesthesis;" the "I move" and "I do" of human existence (1970, 106). However, the physical body differs from the living body [*Körper und Leib*], or the perception of one's body as a subject distinct from others. It is through the living body that a self acts with intention; not solely through the phenomena and the noumena but through the noetic of the life-world (Husserl 1970, 108). We live in "wakeful world-consciousness" and "are constantly active on the basis of our passive having of the world" (Husserl 1970, 108). Here Husserl echoes Kierkegaard's (1980) contention that human existence is driven toward action, as he writes:

> The consciousness of the world, then, is in constant motion; we are conscious of the world always in terms of some object-content or other, in the alteration of the different ways of being conscious (intuitive, nonintuitive, determined, undetermined, etc.) and also in the alteration in affection and action . . . (Husserl 1970, 85)

Human selves negotiate the validity of the world or at least that horizon of the world which they perceive, through their intentional acts as subjects.

Since subjectivity is so intimately tied to consciousness of the world, Husserl (1970) does not discount the possibility that comes from living in wakefulness within this world. As Husserl claims, we live toward those objects and others in existence in such a way as to lend coherency and validity through all actions in the life-world (1970, 144). It is that realm of intermediary, intentional action that facilitates the negotiation of the subject and the world. Husserl (1970) argues, however, that it is essentially the loss of this awareness of intention that constitutes the "crisis" of meaning, particularly within the scientific mind. Individuals will and do act in ways that give rise to habits validating their perception of existence; to a degree, certainty is always needed. However, these habits will be disrupted because we always are living with and affected by others. Here we again encounter that moment of anxiety. It is during these moments of disruption of consciousness that the means of negotiating between subject and object—intention—becomes compromised. One finds oneself compelled to ask the phenomenological question "Toward what should I attend now?" with raw awareness of the possibilities of intention. Husserl would argue that, in order to regain a sense of meaning within these moments, one must engage in radical reflexivity, or the transcendental epochē from the natural attitude into the life world (1970, 148). This consists in living in such a way as to continually investigate the world so that one remains ever attentive to meaning and ontic validity; recognizing that every conscious act implies "an infinite horizon of inactive validities which function with them in flowing mobility" (Husserl 1970, 149). For Husserl, then, the transcendental epochē is not a temporary act but instead a way of engaging the life world in which one adopts an "attitude *above* the universal conscious life . . . through which the world is 'there' for those naïvely absorbed in ongoing life, unquestionably present" (1970, 150). Freedom from anxious moments, for Husserl (1970), comes from the radical, reflective regaining of conscious meaning in all action.

Humankind lives within an infinite number of actual and possible world experiences; our intentional subjectivity makes this so. In those moments of incredible possibility—those moments that demand our attention within intention—we become unsettled. Husserl's (1970) transcendental epochē calls individuals to take an "attitude above" the "pregiven world" and engage that through which "the world *is* at all"—our intentions (1970, 152). It is the contention of the author of this essay that Husserl's phenomenological method reminds us of the need for a place to stand or a place to "fall back" into temporal existence, as Kierkegaard (1980) may argue. For one's consciousness of the world is never "pregiven" but instead is a result of a communal unity. We understand the essences of this world and the connections between objects and our perceptions because of a textured history of understanding shaped by others. We are participants of a historical commitment to human meaning. The life world then becomes not a commitment to what this world means to "me" but to the communal understanding of phenomena that originate always already outside of the self (Husserl 1970). This essay argues that we find this space of communal understanding or the place where we stand within language; an imperfect, temporal system of *imitation*, *identification*, and *communion*. From this contention, the essay now turns to imitation.

Imitation as Confession

Living in the temporal "life world" with others means engaging in a participatory subjectivity. Anxiety manifests not from the banality of passive existence but from the consciousness of possibility of choice. The notions of imitation and order become important here as this essay explores how language creates both the conditions for and the means by which to navigate anxious, reflective existence. St. Augustine's theory of subjectivity, as is illustrated in *On the Trinity* (2002) and *Confessions* (1960), points to the linguistic nature of human temporal existence. Augustine (1960) claims that the individual self is comprised of a series of trinities of action and substance, reflecting our derivative nature as creatures of God, the Trinity. Knowledge, will, and love comprise one of these trinities of subjectivity. It is from this trinity that epistemological and linguistic imitation emerges as key to understanding how the self engages the world of possibility and choice. First, this essay includes a brief synopsis of Augustine's epistemological assumptions.

Augustine's epistemology is propelled by his understanding of knowledge as inseparable from love and will. In Book 9 of *On the Trinity*, Augustine (2002) claims that the mind is equal to the will, which is the same substance or spirit as love. He writes that love and knowledge are not in the mind as a subject, "but they are also there substantially as the mind itself is" (Augustine 2002, 9.4.5). Mind, love and knowledge, in other words, form a trinity of substance from which a self knows. In very simple terms, a person wills to truly know that which he or she loves. If one does not love, one cannot truly "know" an object or an other. This knowledge does not manifest in a possessive love of how an object or other "ought" to be but instead in the willful engagement of how the world "is" as a created order. Augustine (2002) claims that a desire exists within each person to know (i.e., to love) the form and order of God. He writes:

> the more a thing is known, but not fully known, the more the mind desires to know the rest . . . For that form touches the mind that knows and thinks; it reveals the beauty of minds that have been brought together in fellowship by listening to and answering questions through signs that are known (Augustine 2002, 10.1.2).

Augustine (1960) presupposes that every heart, mind and will in God's creation calls out in a restless state to know God and, consequently, to know how to call back to God from within the self. This is a distinctly linguistic process, not only verbal but aural; it is both a questioning and a listening (Augustine 1960; 2002).

According to Augustine, to know is to love, and to love is to will to imitate. Imitation, in this sense, can be regarded a mimetic process of ordering. Augustine (2002) understands imitation as the act of the will in moving toward a good. To move toward a good, or in phenomenological terms—to act with intention toward an object or other—is to engage in a process of ordering. The created world, including the self, has a distinct order toward which one must attend. For Augustine (1960), this is the two-fold order of charity, or first to love God and then to love your neighbor. However, we live in the possibility of disorder. To know disorder is to love out of order, or to imitate a good not true to the self as a created being. This disordering of the self is a distortion of intention; there is always the temptation that, somewhere from the Word of God to the words of our derivative selves, there can be a disconnect of will as the self turns to *a* good instead of *the* Good. As one can imagine, the possibility in a turning causes restlessness, and the heart does not rest until it once again finds rest in God (1960).

With this emphasis on order and imitation, Augustine's (2002) theory of knowledge offers a participatory view of subjectivity. While the self is completely present to itself at all times, one never belongs completely to oneself. To truly know oneself as a self is to know that one is derivative—a creation—and, hence, shaped by reflection and action. After his conversion experience, Augustine (1960) writes in the *Confessions* that he continued to pursue the question of the self. He asks "Who am I, and what am I?" and claims that the answer lies in ". . . not to will what I willed and to will what you willed" (Augustine 1960, 9.1.1). For Augustine, knowledge is an issue of turning; the turning of the will toward God's will, or the participation in and imitation of the created order. Although Augustine (2002) begins with the presupposition that all minds have the capacity to know themselves, he claims that one knows that one has a soul and a mind mostly through participation with a world with others. In the movement of other bodies, a person perceives something like his or her own soul and mind. Augustine uses the example of picking an apple off of a tree. If he picks an apple off of a tree because he desires the fruit and then sees another person picking an apple off of that same tree, he can assume that the other person has a soul just like he does (Augustine 2002, 8.6.9). In other words, to see other beings participate in the world validates the intentions and actions of the self as a self.

This participatory self not only presupposes the existence of others but, ultimately, the existence of a "third." To function without the recognition of the "third," or God, according to Augustine (1960), would cause a self to know and to love out of order. Yet there exists a tension within this relationship of which Augustine (1960) is keenly aware. Although God is always already present to a self, the mind cannot ever hope to fully understand, through its state of temporality, his incommensurability and eternity. This is due to the necessarily *discursive* nature of knowing. According to Eugene Vance (1982) in "Saint Augustine: Language as Temporality," Augustine believes that an incommensurability exists between conventional language and God. God's eternal Word, unlike human language, produces without division of images or signs. Our corporeal nature, however, necessitates that human knowledge of God, of the self and of the world remains divided in a succession of utterances (Vance 1982). These acts of speech, as they reflect images recalled within memory, function as a contract, or a "pact," between the verbal sign and the image of that object as it is perceived in its form (Vance 1982, 25). Yet, since the Fall in the Garden, all selves and their signs are embedded in history and, consequently, only partially and opaquely understood as well as enacted. Akin to Kierkegaard's (1980) understanding of human nature, Augustine (2002) holds that participation in the world and with God, then, is necessarily a linguistic act because humans are temporal creatures. Subjectivity is a ground for discourse among the self and the images of what it knows. What guides the focus of attention of the mind and will, then, ultimately is the Word calling from within and from without of the self. God's Word, as it reverberates from the created world, ". . . speaks to all, but only those understand who compare its voice taken in from outside with the truth within them" (Augustine 1960, 10.6.10). All of creation, including the subjective self, knows by way of words because of the Word.

Confession, then, can be considered the word of truth that is called out of the self by God and spoken forth into the world. Loving God, or knowing God, allows for the word of the self to be begotten or incarnated into utterances. These utterances are an expression of the mind and are therefore equal to the mind. However, once an utterance is spoken into existence it is no longer embodied by the speaker. The utterance becomes wholly separate yet still wholly unified, in love, to the speaker and to God, who called forth the word. Here we turn back to Kierkegaard's (1980) understanding of anxiety as a state of ignorance and of possibility. In a moment of anxiety, or a moment in which we must "speak out" our ethical choice, we imitate a good whose nature we will never completely "know"—we participate with a "third" that manifests as so indistinguishable that we do not see boundaries but only believe in possibilities. At times, these possibilities become too much to bear; we may lose the courage to reflect upon and listen toward what the life-world is calling out for us to turn. Augustine's recommendation for this moment is to recognize that we are creatures who are bound by God and by others. To attend to this truth of Being is to act in love. This love—simultaneously an act of participation as well as an act of rest—is the act of attending to the One found within many. Accordingly, this essay now turns from imitation to identification, or the acceptance of a linguistic, and hence ethical, system as a "third."

Identification as Piety

Kenneth Burke's view of what constitutes human linguistic and ethical systems maintains its origins in both biological and symbolic presuppositions. Burke's (1984) aim, especially in *Permanence and Change: An Anatomy of Purpose*, is to reinstate the metaphor of purpose in discussions of human motivation. His understanding of purpose as is manifested in human symbol systems is greatly influenced by the works of Aristotle and Spinoza, especially in regards to the metabiological view he proposes shapes human existence. In the context of this essay, Burke's (1984) understanding of metabiological identification and linguistic piety are explored as realms in which conditions for anxiety are both formed and navigated, for it is the tension between the biological and the symbolic that allows for human purpose to emerge.

The metabiology of Kenneth Burke views human existence as ecological, or existence as an intersection of both individual motivations and contextual constraints. The internal and the external are continually negotiated within human biological and symbolic systems; we seek balance grounded in existence. Burke identifies this ground as "nature as itself containing the principle of speech" (Thames 1999, 25). Human being has both biological and linguistic attributes that spring from the universal being, or Nature. Like Spinoza's *Deus sive Natura* (God or Nature), Burke's Nature suggests a scene for the human agent that is both finite and infinite (1969, 138). Ecology, then, means for Burke that "Nature is a living organism which manifests a level of organization above that on which separate organisms interact" (Thames 1999, 26). In other words, there exists a definite pattern of relations of organisms within the human ecosystem. There is an order and purpose to the world that human existence struggles to project from the internal to the external; from the finite (Nature) to the infinite (language). As Augustine (1960) would argue, we speak within temporality because we *must*, because we *have first been spoken to* by God.

In his emphasis on ecological order, Burke understands the purpose of the human organism as reflecting Aristotle's notion of entelechy. The human organism engages in both action and motion—the symbolic and the biological—in order to live out its purpose. Because internal realities are merged with external circumstances, the human organism has an affinity for the "proper order" of things from a very young age (Muir 1999, 39). In other words, human organisms, through their understanding of the symbolic, develop a sense of what properly goes with what and how to pursue purpose in their lives. People develop orientations that are organic in that they are biologically grounded and emerge from the essential purposes and gratifications of Nature (Thames 1999). It is the Aristotelian entelechy of human existence to live as bodies that learn language (Muir 1999). However, Burke would also argue that humans have an entelechial motivation to "perfect" our symbol systems. Human existence is "rotten with perfection" in its attempt to find the exact symbolic representative of a non-symbolic world (Muir 1999, 44). Human existence is grounded in the organic yet reaches beyond into the symbolic.

The two worlds of the symbolic and the non-symbolic interpenetrate one another. As a result, humans impose a system of naming on the world and create hierarchy. Here Burke draws on Spinoza's theories as they relate to the metaphor of determination. In *A Grammar of Motives*, Burke describes how we "determine" our world through language:

> And the world of our everyday finite experience, the world of *positive* things like apples, houses, people, is in the Spinozistic vocabulary a world of "negations," because each such positive thing is *determined*, and the determined is that which has its boundaries marked by other things, in brief by things that this particular thing is *not*. Hence his formula, "all determination is negation." (1969, 143)

Burke agrees with Spinoza's notion of determination with his metaphor of the negative. Burke (1984) would argue that humans are in constant struggle to name correctly due to our intrinsic sense of "how things hang together" in the world. He claims that nothing "negative" exists in the nature; there are no such things as a "thou-shalt-nots" in existence except as they are created symbolically by language (Burke 1970, 101). In other words, the names that humans attribute to things in existence not only articulate what they are but also necessarily imply what they are not. Declaring every "yes" means simultaneously declaring a "no," and it is the possibility of misnaming or misrepresenting—picking the "wrong" "yes" or affirming the "wrong" "no"—that lends our intentions a linguistic anxiety. In fact, having to name at all reminds us of our imperfection and the imperfect symbols that motivate and frame our intentions. We are continually frustrated by the only system that through which we know—language.

As we attempt to categorize and define our existence, hierarchy emerges as a symbolic sense of order. Human motivation, then, becomes a subdivision of a larger frame of meaning as is articulated within both the "positive" Nature and the "negative" hierarchy. Burke writes "since we characterize a situation with reference to our general scheme of meanings, it is clear how motives, as shorthand words for situations, are assigned with reference to our orientation in general" (1984, 31). Humans are motivated by the symbolic meanings we attribute to a situation and its named position within the hierarchy of our symbol system. We pick apart Nature—which is whole and continuous in itself—only to put it back together again in a way that reflects the utterances of our symbol system. Simply put, we impose boundaries on existence for the sake of binding ourselves. Human understanding and purpose, then, depend upon the substance of the motive within which they are framed.

The act of framing human motivation is an anxious one. It constitutes reflecting upon the "third" of existence or what Burke might call "the range of piety" (1984, 71). It is to consider that which is at once above and beyond the individual and the collective yet grounded within both. Piety, in Burke's terms, refers to "a system-builder, a desire to round things out, to fit experience together into a unified whole. Piety is *the sense of what properly goes with what* . . . a response which extends through all the texture of our lives" (1984, 74–75). Piety, then, involves a distinctly ethical impulse; it is the consciousness of one's choices as they relate to the "sources of our being" (Burke 1984, 71). The symbolic is always already pulled by Nature, and the biological always already pushes toward the symbolic; like the movement of the muscles of one's jaws and tongue as one speaks, action and motion must coexist. For Burke, it is this dialectic of the biological and symbolic, or the effort to negotiate the tension of symbol-use grounded in a non-symbolic reality, that drives us to continually search for "what properly goes with what," or what may be understood as searching for the "third" realm of possibility (1984, 74). This searching is what Burke deems as our "vocation" when he writes "our calling has its roots in the biological, and our biological demands are clearly implicit in the universal texture. To live is to have a vocation, and to have a vocation is to have an ethics or scheme of values . . . " (1984, 256). Piety acts as a transitional stage between orientations as "new meanings" are deemed congruous or incongruous to motivating frameworks (Burke 1984, 69). Piety is the opening of possibility for interpretation.

The discussion of anxiety here emerges as one considers how our symbolic systems are inherently ethical. Like Husserl (1970), Burke (1984) contends that our actions are always intentional; our consciousness of the world is value-laden due to our historical and social nature as human beings. Burke himself is keenly aware of the various freedoms that emerge from linguistic possibility. In a footnote in the chapter on "Recalcitrance" in *Permanence and Change*, Burke writes:

> Once you introduce a point of view into the universe, however, (as is introduced by biological vocation), a new factor enters. The point of view requires an interpretation of events, a reading of the recalcitrant factors favorable and unfavorable to the point of view. But an interpretation can be wrong. Hence, a point of view introduces the possibility of error. But where there is the

> possibility of wrong interpretation, there is also the possibility of a right one. The *freedom to err* argues a *freedom to be right.* (1984, 257)

Like Kierkegaard (1980), Burke (1984) understands that the moment of the "I can" in anxiety is a dizzy moment of intense freedom. It may not be the freedom that we hope for, but it is freedom nonetheless. We are driven back from the innocence (Kierkegaard 1980) with our intentional acts (Husserl 1970) within the symbolic (Burke 1984) only to hope to find communion with the One that constitutes our order (Augustine 1969). We want to identify with the sources of our being, to remain pious to our purpose as human selves. It is an anxious vocation, especially since human desire to commune with the One is one in the same with communing with alterity. With this, this essay concludes by turning to Emmanuel Levinas and the metaphor of communion.

Communion as Temporal Communicative Ground

This linguistic order that we strive for, or the intentional projection of the boundaries of the self within the world, finds further dissonance with the fact that we are selves in a world of others. This has been a contention of all theories of subjectivity and consciousness that have been discussed thus far—one may begin with the self in understanding experience, but one cannot end there. Simply put, living with others makes us anxious. Biologically, we experience detachment from others, both through birth and death. Symbolically, we experience alterity, or an initial disconnect through the recognition of difference. Yet it is a disconnect that we must face because when one engages "separation as life," according to Emmanuel Levinas (1969, 109), one lives in piety to *a priori* social relations of the self.

Levinas describes the life of the self as experiencing a weight of Being. He points to the fundamental characteristic of human existence as the need to escape from Being (Manning 1993). Within Being, the self finds burden. This occurs not because there is a lack of self, but instead the self, in solitude, becomes so present to oneself that it becomes too much to bear. The self in solitude, against the weight of existence, experiences a restless state of *malaise* and even *nausea* as one experiences "the anxious realization that we are thrown into being without anyone having asked us if we wished to be there or having told us where we are going" (Manning 1993, 35). Levinas describes the existence of the self as a raw presence without hope for a future. He writes "one is there, and there is nothing else you can do about it, nothing to add to the fact that we have been brought here entirely, that all is consummated" (Manning 1993, 35). A person, standing by himself or herself, can feel overwhelmed by the responsibility of being brought entirely into existence and standing there at what seems like a great distance to anyone or anything else. With no one or no thing standing by one's side, the boundaries of the self are indeterminable. Levinas (1969) introduces sociality as a necessary means of escaping the constant presence of the self by way of responsibility toward others.

Embracing sociality means engaging alterity. Levinas (1969) presupposes social relations as being the constitutive relation of the self to the world. The Other calls us into a significant relationship with life and with ourselves. It is what will liberate us from the weight of being. Levinas recognizes, however, that at the basic level, the self is alone; one is always already completely separate from everyone else. As he states in *Existence and Existents*, one hears the constant reminder of a "monad": "I am not the other. I am all alone" (Manning 1993, 55). As a result, we do not experience others as "alter egos" of ourselves but instead as Others that are completely different from ourselves (Manning 1993, 52). We can never completely understand the Other; the Other is experienced as complete alterity. However, Levinas contends that being faced with absolute Otherness, or radical alterity (such as death), confirms that life has a plurality to it (Manning 1993, 52). There is something outside of the self, which we cannot completely understand but which we *must* engage. For Levinas, liberation from the weight of being through the engagement of alterity is constituted by both time and the other (Manning 1993, 53). We are called out of the anxious presence of the self by our task, in time, to care for the other.

Time has infinite possibilities due to our social relations with others. Unlike Heidegger, who supports the understanding of temporality through the individual *Dasein*, Levinas contends that time must be interpreted through the essential social relations of human existence, or the relation between the subject and the other (Manning 1993). Manning describes Levinas's understanding of the present as "the instant at which beings emerge from being in general," or that moment in which a self recognizes itself as so (1993, 62). This moment is characterized by a kind of locking of identity, and one becomes aware that one stands, alone, as a separate existent. The only movement that is offered in the present is "an inevitable return to itself" (Manning 1993, 64) and to this moment, never to a new moment. Akin to Kierkegaard's (1980) state of anxious ignorance, the self is essentially born in innocence; one cannot know anything outside of the self, and, hence, does not really understand the possibilities of the self. Levinas's contention is that only the Other offers newness and continuity to life. Sociality offers a *new* beginning, a future that exceeds the parameters of what is foreseeable and possible in the presence of solitude. According to Levinas (1969), the future is what should characterize temporal existence–a future driven not by death or limitations, but by love.

Love, or *eros*, plays an important role in Levinas's understanding of the possibilities that manifest in the temporal. In the same vein as Augustine (2002), Levinas understands love as a matter of the will or of intention; it is a turning toward the other *as* other, not as the self. This is characterized most poignantly within the lover-beloved relationship, because "in *eros* the Other gives to the subject a future that connects to the subject's present" (Manning 1993, 74). Levinas claims that it is within the *eros* relationship that the alterity of the Other is experienced with the most intensity; in love one has an inexhaustibly intense desire for union (Manning 1993). However, love presupposes that one cannot possess the Other. Instead, one honors the fact that the beloved remains in otherness. Hence, the lovers live unsatisfied—within each other's arms but never completely within reach. According to Levinas, however, this is an occasion for joy, for the social relation, especially love, "opens up to the subject its absolute future, its own future as mystery and surprise" (1993, 77). Time becomes fruitful; it always recommences and is always born anew. Time offers pardon as "it flows from the other to the subject and releases the subject from the weight of its present" (Manning 1993, 78) and into infinity.

With this notion of alterity, the self experiences others not just in relation to the self, but instead one must engage the experience of the other with a radical reflexivity—we know most completely when the Other before us helps sediment our perceptions. Attending to alterity, Levinas does not claim that social relations offer us answers. Instead, he contends that they allow us to ask the right questions. The question for Levinas (1985), as he discussed in *Ethics and Infinity*, is not one of one trying to figure out how Being unfolds in human existence. Questions such as "what is Being?" or "what is human being?" are not *a priori* in our understanding of what it means to be human. Instead, in sociality we find ourselves asking "Do I have the *right* to be?" (Levinas 1985, 121). In solitude, how can we ever ask that—or even hear an answer? The Other binds us in meaning, and Levinas (1985) claims that these are boundaries that were determined *before* human language, *before* metaphysics, *before* the self. These are discursive boundaries of the Saying, for it is a command that " before the face [of the Other] I do not simply remain there contemplating it, I respond to it" (Levinas 1985, 88). It is this ability to ask within the Said "Do I have the *right* to be?" that one hears the Saying answer "yes." As beings in the world of others, we hear an affirmation of purpose in determined life together. The metaphor of communion points to a metaphysics of responsibility—the essence of all things is one of proximity (Levinas 1985). We are entirely separate from all other beings in this world, but we are not alone in Being.

We ask restless questions of our life as selves and ask even more anxious questions of our life with others, yet we are selves in a world of others. The ground that we discover out of anxiety becomes a ground of communion—or, in other words, a communicative ground. It is existentially driven and shaped through language. Here we see the culmination of all of our metaphors in understanding anxiety as a condition of our existence as language-using selves in a world of others. We intend, we attend, we imitate, and we identify meaning—biological, symbolic, and spiritual meaning—in order to bind the self. As was mentioned above, sometimes these boundaries are misnamed, misrepresented, misused, or missed completely. Yet, we need them. We need them in order to affirm where we stand in the face of uncertainty, when we're faced with losing our ground. Accordingly, this essay now concludes with a brief reflection on anxiety and ethical choice.

The Ethic of Anxiety

Understanding anxiety from a phenomenological perspective moves one out of the realm of therapeutic renderings of and techniques for reducing stress and panic ("I have anxiety; now how do I get rid of it?") and into a philosophical and ethical engagement of attention and questions ("I am anxious; now how should I live?"). To engage anxiety as an ethical force points to the embedded nature of our existence. Anxiety, as Kierkegaard (1980) understands it, presupposes our existence as creatures embedded in the temporal. Yet we experience momentary spells of "dizziness" as the world of possibility sweeps the ground from under our feet (Kierkegaard 1980, 92). Husserl (1970) understands our nature as embedded in intentional consciousness, sedimenting our perceptions of the world through history with others. Yet we experience a transcendental epochē in which we stand above the normative renderings of unreflective existence and find freedom in radical reflexivity. Augustine (2002) understands that our existence is bound by God and by others in imitative, discursive embeddedness. Yet we find rest in God and in confession of the order of the self. Burke (1984) understands the confession of symbol systems as the range of our piety; we do not find order in creation but instead create order. Yet we find purpose by embracing the only freedom which our linguistic existence allows—the freedom either to be right or to be wrong accordingly to systems of our own making. Finally, Levinas (1985) understands ethics as an infinite task of our life with Others. Yet in the love of the Other we find endless possibilities of the self in relation.

It is clear that each of these thinkers view anxiety as a product of existence—but not just existence as it *is* but existence as we *believe it to be* through the use of our symbol systems. They may offer glimpses of boundless possibility of the self, but, each in his own way, they contend that we must "fall" back into temporal existence together, for that is where we belong. Anxiety as possibility, then, necessitates a realm in which to live out that possibility, for ethics is a practice—an attending, an imitating, an identifying, a communing—toward and with a given good. If we are lucky, we find ourselves embedded in an existence which brings "crisis," for with crisis comes reflection of choice and, more importantly, the affirmation of purpose.

REFERENCES

Saint Augustine. 1960. *The Confessions of St. Augustine*. Trans. J. K. Ryan. Garden City, NY: Image Books.

———. 2002. *On the Trinity (Books 8–15)*. Trans. Stephen McKenna. Cambridge, MA: Cambridge University Press.

Burke, Kenneth. 1969. *A Grammar of Motives*. Berkeley; Los Angeles; London: University of California Press.

———. 1970. *Rhetoric of Religion: Studies in Logology*. Berkeley; Los Angeles; London: University of California Press.

———. 1984. *Permanence and Change: An Anatomy of Purpose*. Berkeley; Los Angeles; London: University of California Press.

Foucault, Michel. 1970. *The Order of Things: An Archaeology of the Human Sciences*. New York: Random House.

Heidegger, Martin. 1962. *Being and Time*. Trans. J. Macquarrie and E. Robinson. San Francisco: Harper San Francisco.

———. 1992. "Letter on Humanism." In *Basic Writings from Being and Time (1927) to The Task of Thinking (1964)*. Ed. David Farrell Krell, 213–266. San Francisco: Harper San Francisco.

Husserl, Edmund. 1970. *The Crisis of European Sciences and Transcendental Phenomenology*. Evanston, IL: Northwestern University Press.

Kierkegaard, Søren. 1957. *The Concept of Dread*. Trans. Walter Lowrie. Princeton, NJ: Princeton University Press.

———. 1980. *The Concept of Anxiety: A Simple Psychologically Orienting Deliberation on the Dogmatic Issue of Hereditary Sin*. Ed. and trans. Reidar Thomte. Princeton, NJ: Princeton University Press.

Levinas, Emmanuel. 1969. *Totality and Infinity: An Essay on Exteriority*. Trans. Alphonso Lingis. Pittsburgh: Duquesne University Press.

———. 1985. *Ethics and Infinity: Conversations with Philippe Nemo*. Trans. Richard A. Cohen. Pittsburgh: Duquesne University Press.

Manning, Robert John Sheffler. 1993. *Interpreting Otherwise than Heidegger: Emanuel Levinas's Ethics as First Philosophy*. Pittsburgh: Duquesne University Press.

May, Rollo. 1977. *The Meaning of Anxiety*. Rev. ed. New York: W.W. Norton & Co. Inc.

Muir, Star A. 1999. "Toward an Ecology of Language." In *Kenneth Burke and the 21st century,* ed. Bernard L. Brock, 35–70. Albany, N.Y.: State University of New York Press.

Schrag, Calvin O. 1986. *Communicative Praxis and the Space of Subjectivity*. Bloomington, IN: Indiana University Press.

———. 1997. *The Self after Postmodernity*. New Haven, CT; London: Yale University Press.

Thames, Richard H. 1999." Nature's Physician: The Metabiology of Kenneth Burke." In *Kenneth Burke and the 21st century,* ed. Bernard L. Brock, 19–34. Albany, N.Y.: State University of New York Press.

Vance, Eugene. 1982. "Saint Augustine: Language as Temporality." In *Mimesis: From Mirror to Method, Augustine to Descartes,* eds. John D. Lyons and Stephen G. Nichols, Jr., 20–35. Hanover, N.H. and London: University Press of New England.

Part Two

Response-ability as Inner Transparency

Interpersonal Crisis Communication in the Workplace

Professional Civility as Ethical Response to Problematic Interactions

Janie M. Harden Fritz
Duquesne University

INTRODUCTION

A growing body of research suggests that today's workplace faces issues of incivility (Andersson and Pearson 1999; Pearson and Porath 2005; Zauderer 2002), social undermining (Gant, et al. 1993), workplace aggression (Baron and Neuman 1998; Coombs and Holladay 2004), workplace deviance (Bennett and Robinson 2000), bullying (e.g., Namie and Namie 2000; Tehrani 2001), employee emotional abuse (Lutven-Sandvik 2003), antisocial behavior (Giacalone and Greenberg 1997), and other types of problematic work misbehavior (Vardi and Weitz 2004). These interpersonal phenomena yield high costs for organizations and persons, resulting in stress and burnout (Omdahl and Fritz 2006), absenteeism (Kimura 2003), and turnover (Chapman 1993). At the same time, signs of hope emerge through research on forgiveness in the workplace (Metts, Cupach, and Lippert 2006), civility (Sypher 2004), behavioral integrity (Simons 2002), and of how the "relational architecture" of jobs can encourage people to make a positive difference in the lives of other employees (Grant 2007).

Arnett (2006) and Arnett and Fritz (2003, 2001) articulated the metaphor of professional civility as a professional communicative ethic for organizational life, understanding organizations as public places. Arendt's critique of the blurring of public and private into the realm of the social, posited by Arnett (2006) as an essential foundation for professional civility, is an important component of this "fuzzy metaphor" (Arnett 2001) of professional civility. In this paper, organizations are reconceptualized as civic space, distinguishing the realm of the civic from the realms of public, private, and social. Against this backdrop, we offer professional civility as a style of communicative and rhetorical practice designed to combat the interpersonal communicative crisis of incivility in the workplace. The first part of the essay reviews literature on the concept of professions and of civility. The second part develops the idea of professional civility as articulated in previous work by Arnett (2006) and Arnett and Fritz (2001, 2003). The third part develops the idea of organizations as civic space. The fourth part describes performative elements of the communicative style of professional civility with an eye to both a potential operationalization of the concept and further conceptual and philosophical development for identifying its appropriateness for organizations as civic space.

Professionals and Professionalism

According to Kimball (1995), "Ideally, a 'profession' is a dignified occupation espousing an ethic of service, organized into an association, and practicing functional science" (17). Professionalism stems from idealizing an occupation and involves honor for the profession (2). The word "profession" has its roots in the word "profess." To profess means to speak forth, to proclaim, to stand in the place of. In the middle ages, guilds were groups that formed to protect their knowledge and abilities from others. They were concerned that the image of all would be affected by unskilled practitioners proclaiming to be one of them. We see a similar urge today as professions seek to control by licensing those who have the right to claim the title, or the identity, of a profession.

The terms "profession" and "professional" have a rich history. Kimball (1995) traces the use of these words up to the articulation of "the true professional ideal" in the United States in 1894. He notes six rhetorical moments during which the term developed: (1) the extension of "profession" from referring to a religious vow to denoting the group who made the vow; (2) the shift from reference to clergy to dignified, non-religious occupation; (3) the introduction of terms "learned professions" and "liberal professions," which identified certain dignified occupations; (4) the displacement of "professed" by "professional" as the adjective denoting "occupational" or "vocational"; (5) the introduction of noun "professional" to replace "professor," which was narrowed to the field of education; and (6) the sloughing away of terms "learned professions" and "liberal professions" in the early twentieth century, as "professional" began to refer to many vocations, just as it had in eighteenth-century English (15–16).

Early in the twentieth century, the time Kimball's story ends, Sullivan's (1995) story of professionalism begins. The ideal of professionalism articulated by reformers in Great Britain and the United States pictured highly skilled workers with a socially responsible sense of calling, bringing to occupations an almost religious dedication combined with civic ideals "respecting diverse viewpoints, while aiming at a consensus about common purposes" (1995, xvi). "Professionalism is a very loosely defined term," originally referring "to the classic honorific occupations of medicine, the bar, and the clergy" (2). Occupations classified as professions are typically characterized "by three features: specialized training in a field of codified knowledge usually acquired by formal education and apprenticeship, public recognition of a certain autonomy on the part of the community of practitioners to regulate their own standards of practice, and a commitment to provide service to the public which goes beyond the economic welfare of the practitioners" (2).

Professionalism "has emerged as a widely esteemed and sought-after virtue" (Sullivan 1995, 3). Professional behavior is an implicitly agreed upon standard: "few speak of professional behavior as existing only in the eye of the beholder" (3), and charges of "unprofessionalism" seem to be generally understood. Professional work "requires the practitioner to adhere to demanding standards of competence and public service" (4). Professions hold a "special dedication and clear accountability which, to common sense, distinguishes a profession from trades and businesses. A profession is 'in business' for the common good as well as for the good of its members, or it is not a profession" (5).

The "good" embraced by professions include those of useful labor in service to others and the resulting common good to the community that these goods entail. Professionals are now part of large organizations, and they are also pressured by economic forces to compete and to give in to profit when professional standards of excellence come into conflict with economic goals (Sullivan 1995, 6). One example of the expectation that professionals will serve the public good is work done *pro bono publico*. This giving to the public good is "the ethical spirit of civic life manifested and given specific direction in professional life: the spirit of profession as a calling (10).

Civility

An examination of the concept "civility" should be preceded with a brief examination of the antecedent term "civil society," a concept with a long tradition (Kumar 1993). Kumar traces the use of the concept "civil society." Until the end of the eighteenth century, the term *civil society* was synonymous with the state or political society. Civil society referred to "a social order of citizenship where men . . . regulate their relationships and settle their disputes according to a system of laws; where 'civility' reigns, and citizens take an active part in public life" (376–77). At some point civil society came to refer to something other than the state. Kumar credits American and British thought for this divergence: the writings of Locke, Paine, Ferguson, and Smith elaborated a sphere of society distinct from the state and with forms and principles of its own. "Civil society is a part of ethical life, the part that provides the middle term between the family and the state" (Kumar 1993, 378). Civil society's domain encompasses the life of community as a whole (378), echoing Gramsci's broader definition: "Civil society is the sphere of culture in the broadest sense. It is concerned with the manners and mores of society, with the way people live. It is where values and meaning are established, where they are debated, contested and changed" (382).

Kumar (1993) observes that the conception of the person within the context of the term civil society transcends the individual's sphere. "The 'concrete person' of civil society differs from the isolated subject of the sphere of morality in that he gradually comes to recognize himself as a member of society and realizes that to attain his ends he must work with and through others" (378). He cites de Tocqueville's concept of "political society," consisting of political and civil associations, supplies the independent eye of society that exercises surveillance over its public life, "educates us for politics, tempers our passions and curbs the unmitigated pursuit of private self-interest" (381). Here one can discern a disjunction between private morality and public ethics stemming from the term "civil society," to which the concept of civility is closely tied. Bryant (1993), in his critique of Kumar's (1993) treatment of the concept "civil society," suggests that civility is "the equable treatment of others as fellow citizens however different their interest and sensibilities" (Bryant 1993, 399). Civility, he suggests, should not be equated with liking or affection: "Civility, it should be noted, is a cool concept. It does not require us to like those with whom we deal civilly, and as such it contrasts strongly with the warmth of communal, religious, and national enthusiasms" (399). As further elaboration on the concept of civility, Bryant cites Nancy Fraser's development of Habermas's public sphere in which communication and discourse are vital: " . . . [C]ivil society refers to social relations and communications between citizens." Those interacting are in the role of citizen, with ascribed and avowed identities (Collier 2003) as citizens.

If "civil society" carries ethical meaning, then to what extent can civility be considered an ethical quality? Walzer (1974) suggests that civility refers to "a certain decorum of behavior" (597) and consists of the "moral and political qualities that make a good citizen," once known as "civil righteousness" (593). More recently, he observes, civility has come to connote social virtues of orderliness, politeness, and seemliness, which is oriented toward the private, rather than the public, realm, and thereby, it would seem, oriented away from ethics into the sphere of manners.

Yet civility's sense still resides in the public realm; Walzer (1974) suggests that it can refer to good behavior and everyday law-abidance. There is also more connoted in the term "civility" than overt behavior; it seems to refer to the attitude or spirit in which one complies with law or follows social rules (598).

Interestingly, civility's social connotation of manners may have lessened the effectiveness of citizen participation in public affairs. Walzer (1974) notes this effect in his discussion of the opposition of civility and intense public participation: "The improvement in social discipline seems to have been accompanied by a decline in political passion (600), for political passion means disagreement, argument, taking sides. Civility, with its shift in meaning toward socially polite behavior, has drifted in meaning toward the notion of tolerant relations—the type of "friendship" that Aristotle said characterizes members of the same political community. "Tolerance is a crucial form of civility . . . The more tolerance, the less political involvement, because political involvement creates strife" (601). Civility, then, seems to connote in this sense public "niceness," or courtesy, generosity, and a concern for rules (610). Civility and civic virtue are at odds because one is predicated on tolerant acceptance and the other on standing for something.

Becker and Horowitz (1970) call San Francisco "the culture of civility," focusing on civility as tolerance. Deviance can be a civic resource. However, there are limits to tolerance; behaviors and contexts that cause violence are avoided. Therefore, civility is not completely tolerant, but only of those aspects of behavior that do not threaten civic life. The city's public view of itself guides civil behavior, regardless of private feelings of individuals. However, Becker and Horowitz note that civility seems predicated on sufficient resources; if economics becomes a question, there is less tolerance.

Wood and Bloch (1995), basing their concept of civility on Habermas's (1992) public sphere as well, also consider civility as tolerance: " . . . [C]ivility [is the] enlightened tolerance of diverse viewpoints and ways of life" (124). However, they conclude from their study of discussion of a controversial issue in the United Methodist Church that decision rules and procedures and common beliefs, traditions, and loyalties generated civility despite a controversial issue. Therefore, the ability to have a disagreement, to grapple with contending ideas, and still exhibit civility may require a common ground or a goal. Civility, then, need not be lost completely to social acceptance.

Barrett's (1991) discussion of rhetoric and civility shows us how one may be both passionate about issues and respectful of others. He ties civility and rhetoric to an ability to live in common with other human beings. "To engage rhetorically is to affect others strategically with acquired roles . . . necessarily constrained by their rhetorically behaving selves . . . each acting in their own interests. This is a persuasive enterprise of mutual interest and self-interest . . . it is a civil and dialectical process of social interaction: society's most humane means of implementing essential checks and balances on behavior" (1991, 10). Civility, therefore, does not merely serve the self; to be civil is to have some criterion other than the self as judge of behavior. "It is through rhetoric that culture is sustained" (11). Rhetorical civility is therefore a social good, an ethical construct manifested in communication: "Civility is a social good or virtue—an ethical value—" that is expressed rhetorically (11).

Barrett (1991) cites many theorists who focus on the importance of civility to the public order. From Freud: "Civility arose as an expedient accommodation from the human struggle between individuality and the claims of the group" (Barrett, 146). He cites Buber: "Actual humanity—civility . . .—exists only where unfolds the capacity to confirm others appropriately" (Barrett, 147). According to Burton Zwiebach: "A dimension of the 'common life' characterized by replacement of 'intemperate self-interest with concern for our fellows and with traditions of restraint, forbearance, grace, tolerance, compassion'" (Barrett, 147) Lawrence Rosenfield and colleagues suggest three criteria of "'man at his best'" (qtd. in Barrett, 147): The civil person is thoughtful about self and others: reflective regarding what is said and receptive to others and personal responsibility; careful: respecting, responding, empathizing; and exhibiting good humor and flexibility. From Heinz Eulau, we see that civil behavior is contrasted with self-centeredness and taking life's disappointments as personal affronts—for example, blaming others, resentment, and throwing temper tantrums (Barrett 147). Individual persons must recognize that they play an important role in making society work. Each person has responsibilities to others.

Barrett (1991) finally explicates his own definition of civility, consisting of eight elements. First is knowledge, awareness of others and self, aware of right and wrong choice making. Caring about others is basic to achieving one's own ends. Second is will—the civil person is cooperative and motivated to help shape public order. Third is respect for self and others. Respect requires restraint and a tolerance for the chosen roles of others and the rules of the community that keep the community intact. Fourth is the courage to take a risk, to hear another's message that may threaten oneself. Fifth is ability: rhetorical acumen—perceiving reality, predicting outcomes, balancing self-interest with community welfare, similar to the interpretive competence of Trenholm and Jensen's (1996) competence model. Sixth is independence, standing apart. One must exhibit autonomy of being when necessary or a healthy separateness. When one is not separate from another, no real choice regarding relating can be made. Seventh is freedom: Values related to civil liberties—appreciating the democratic process; this aspect of civility is culture specific. Finally, civility includes responsibility, being accountable for one's actions. This attribute reflects social regard and good treatment of others.

Civility is related to, but not identical to, politeness. From Craig, Tracy, and Spisak (1986) we gather assumptions of politeness theory. All persons are concerned about the public self-image/identity they present to others, known as face (Goffman 1958, 1967, 1971). As originally conceived, there are two kinds of face: negative and positive (Brown and Levinson 1978). Negative face is the desire not to be impeded or put upon or constrained or restricted by others, or to be free in one's actions. Positive face is the desire to have others approve of and appreciate one's self, personality, desires, and wants. Face is a socially created product/process that can be lost, maintained, or enhanced and must be attended to in interaction. We must all work together to maintain each other's face. People, being rational, will attend to meeting their goals efficiently; therefore, one will meet the face needs of others in order to have one's own face needs met. Certain communicative acts, such as requests, threats, criticism, threaten face. The degree of threat is influenced by social distance between conversants, power or status of one relative to the other, and how much of an imposition the act represents. We meet face needs through politeness strategies, which vary in their redress to face needs. For example, a positive face strategy is one that shows the other that the other's qualities are appreciated. A negative face strategy recognizes the other's need to be free and unimpeded.

Politeness and civility are sometimes conflated with the concept of manners. Judith Martin (1982) suggests, citing Erasmus, that "It is part of the highest civility if, while never erring yourself, you ignore the errors of others." To Miss Manners, politeness is not an ethical issue but one of appropriate behavior: "As your conscience might trouble you if you do anything immoral, your sense of embarrassment should be activated if you do anything unmannerly" (1982, 10). "Manners involve the appearance of things, rather than the total reality" (13).

Civility is a deeper concept than politeness. Its goal is to enable the larger social whole to function. Civility requires an attitude of willingness to participate and involves procedural adherence. Civility, for some, involves a moral quality, where politeness does not (in either the academic or popular uses of the term). Both politeness and civility involve a move toward predictability, toward order and stability in personal and public life. Both are social in nature, but one is directed toward the navigation of social life for personal face maintenance; the other is directed to the maintenance of corporate life for the corporate good. Politeness is focused on individual goals and operates conceptually at the micro level, with the relationship as the background.

Civility is focused on collective goals, with the community or collective as the background. Furthermore, politeness theory suggests that politeness functions to present the self as a competent member of society. Civility, in order to promote the good of the whole, may require a suspension of politeness, a focus away from relational elements of the self and toward a task, project, or larger good beyond preservation of feelings.

Barrett's treatment gives a strong link between civility and politeness: people act in their own best interest, according to roles that they avow. Civility is communicative in nature and functions on behalf of civil society and thereby on behalf of all its members. Exhibiting appropriate civility allows persuasion within the bounds of acceptable behavior. A rhetorical disposition would allow one to be as polite as one could be and still make one's point persuasively, thereby meeting the demands of civic life and individual self-presentation goals.

In organizational settings, an interactional ethic of civility among professionals addresses the needs of the larger collective, promoting productive interaction and task focus. The next section addresses professional civility as an organizational communicative ethic.

PROFESSIONAL CIVILITY AS ORGANIZATIONAL COMMUNICATIVE ETHIC

Arnett and Fritz (2001) offered an initial treatment of professional civility as a communicative organizational ethic, situating ethics in an organization's mission (Arnett 1992; Nicotera and Cushman 1992) rather than in the personal preferences of the individual, termed *emotivism* by MacIntyre (1984). That initial definitional statement located professional civility (1) within organizations as public places governed by a clearly articulated mission, or a "local home" (Arnett 1992), which was assumed to provide guidelines for organizational members' judgment and action; (2) within guidelines of a particular profession with identifying norms of conduct; and (3) within everyday understandings of "professional behavior" (e.g., speech and action) shared implicitly by members of the marketplace. This implicit or "common sense" definition was considered to be tied to pragmatic notions of interaction supportive of face needs of others claiming a particular role identity within a profession (Scollon and Scollon 1995), including both distance and connection. The need for "distance" in this context, though not stated explicitly in Arnett and Fritz (2003), can be understood as freedom of movement and definition of self as professional, while "connection" can be understood as recognition of shared and equal membership within an organizational setting or local home. Professionals adhere to the profession's expectations and the norms and culture of a particular organization within which their professional work is situated.

Arnett and Fritz (2003) provided further explication of the metaphor of professional civility as a background framework for communicative praxis within an organization as a public space. Professional civility provides a communicative background uniting word and deed, permitting implementation of an organization's mission, bringing the efforts of all toward a common project. Citing Seligman (1992), who recognizes the need to reclaim public space by protecting the public and the private domains rather than collapsing them, Arnett and Fritz argued that public and private domains interact dialectically but are fundamentally different. They suggest that the public domain offers trust for diverse outsiders by requiring a different form of constraint than does the private domain.

Arnett and Fritz's (2003) review of literature noted increased attention to the relationship of public to private life, citing Arnett and Arneson (1999), Rawlins (1998), and Sennett (1976). The historical shift of production to industry from the home as noted by Gurstein (1996), Marks (1994), and Scollon and Scollon (1995) made the home a refuge of private life to which one retreated to avoid the impersonality of the public sphere, creating a dichotomy between public and private life. The post-World War II therapeutic approach to communication, which stressed self-disclosure and authenticity as a preferred style of communication regardless of context (Arnett 1997), created an ideology of openness in interpersonal encounters. Their review highlights the work of Marks (1994), who notes the turn toward privatization or individualization of the public sphere, particularly in organizational life, turning the public world into a private space with the potential for intimacy in which self-disclosure permits entry into one's private world. The remainder of their essay offered guidelines for managers who elected to follow a metaphor of professional civility to support and maintain an institution's mission in a postmodern moment of competing virtue structures, ethics, and types of organizational mission. This further development of the communicative ethic of professional civility continued its initial articulation as requiring attention to both the profession and the local home, adding the task as a third component:

Professionals need to show respect for both the persons with whom they work and the vocation that calls them to work each day. A story that informs one about what it means to be a professional in a given area—whether health care, teaching, or engineering—must be complemented by a story of respect for co-workers, superiors, subordinates, customers/clients/patients, and for the local home, the local organization. Professional civility begins with threefold respect: the craft or profession, persons at work, and the local organization.

In the most recent treatment of the organizational communicative ethic of professional civility, Arnett (2006) draws explicitly on the work of Hannah Arendt to argue against the invasion of the public by the private, focusing on this blurring of public and private into the social as a danger to both public and private life.

Professional civility assumes an orientation to organizations as public places, the importance of coordinated roles, and the need for boundaries protecting public and private life. Professional civility assumes limits for expectations of persons and their employing organizations: organizations are not designed as places of self-fulfillment, and employees are not designed to give the totality of their lives to organizations, permitting the public to replace or overwhelm the private sphere. Organizations have a responsibility for public articulation of mission or guiding narrative that has fidelity, in which word and deed are connected. Professional civility rejects a human relations model in which organizations attempt to meet individual needs as a means to increase production. Professional civility embraces a human resource model, elevating ideas, creativity, and productivity as the focus of organizational life. Professional civility involves respect for others in the workplace and a willingness to work with those who one may not like. Professional civility suggests following good ideas, from whatever source, whether a colleague, a person with a work style different from one's own, a secretary, or a CEO. Ideas, not the individual, propel professional civility in an organization (Arnett 2006, 234).

The need to differentiate public and private spaces is tied to a number of coordinates, including the need to experience life that is balanced in multiple ways such that work does not bear the sole burden of fulfillment of a life. Organizations, likewise, bear the responsibility of accountability to the public sphere, balancing the need for profit with the need to offer publicly fair and just procedures and protections for employees and communities, not demanding the soul of employees through invasion of their privacy and outside lives. The dialectical tension between person and institution is managed by public statement of expectations for employment, the dignity of organizational roles that permits focus of attention on work rather than the personal attributes of others, permitting others to do their work without interference of those who think it should be done "my" way, discourse that avoids social undermining of others, and avoidance of feigned intimacy with those one does not know well.

Consideration of an ethic of professional civility suggests the need for further development, specifically, to identify the nature of the space of organizational action in a way that permits differentiation of organizations from the sphere of the private and of the public without the danger of the social that Arendt warned against. Organizations do not, in the final analysis, offer the multiplicity possible and desirable in the public sphere. Organizations do have particular missions that guide them, and there must be limits to multiplicity of purpose and public virtue structure if coordinated action is to take place. For this reason, we reconceptualize organizations as civic space—a particular, bounded instantiation of the public that retains a non-private character but does not slip into the domain of the social with its blurring of public and private.

Organizations as Civic Space

The notion of "civic" implies a particular focus or standpoint, a collection of shared practices for coordinated communal activity. The "civic," in this sense, has a different character that the multiplicity of the public realm, highlighting a specific, articulated mission and vision. The civic is not in the realm of the private, for it involves others who may be different from oneself in important ways. The civic becomes a counterpart to Arendt's "social" in that it does not confuse or blur the public and private realms. The "civic" can be considered a subset of the "public," one of the many localities or "petite narrative" formations that define a set of shared goods or purposes as a defining feature of organizations, a community of memory or public dwelling place (Arnett, Fritz, and Bell 2009) that announces a story that gives its members roles and identities. In this realm of the organization as civic, multiplicity and diversity occurs within the parameters of a publicly stated mission, and the horizon of difference works within mission limits.

Conceptualizing organizations as civic spaces permits professional civility as a discourse and behavioral style to avoid Arendt's realm of the "social," a context in which a false sense of intimacy pervades the public sphere and distance is lost. The civic invites, instead, relevant personal information and discourse within bounds to provide a backdrop of human engagement for the shared task of labor that neither feigns intimacy nor eclipses natural human warmth and contact. The "managed smile" (Arnett 1992, 114) of the social, which resonates with Hochschild's (1983) warnings about emotional labor, is replaced by behavior that turns phenomenologically toward lines of action that focus on common tasks, minimal common agreement on a project, discourse that permits space to others and engages conflict in productive ways. The temptation to discourse focused on gossip, personal matters, problems of others, and social comparison that moves public spaces toward this realm of the social is resisted through common agreement to accomplish labor and support from organizational leadership to guide those practices. Professional civility does not eschew relationships that develop organically as a byproduct of shared labor; task-centered action can generate relational connection and a sense of community (e.g., Gerety 2001). Indeed, such relational formation is in need of scholarly examination (e.g., Wood 2006). The restraint required for professional civility in an organizational space understood as "civic" rather than social or public is guided and required by an organizational mission that directs the focus of attention away from individuals and toward roles as persons within an organizational narrative. Professional civility as the "behavioral uniform" of employee conduct permits a degree of variability in enactment but offers a horizon of acceptable action made public through the particularity of a location within a larger sphere of organizational forms and possibilities.

Professional Civility as Rhetorical Communicative Praxis: Foundations for an Operational Definition

In the remaining pages, I offer reflections on professional civility as it might look in practice. Although the material takes a prescriptive tone, it is offered in that fashion as a starting point for an operational definition. In order to provide the widest possible engagement of the term, approaches ranging from the philosophical to the quantitative may offer insights into this potentially complex concept. I end by highlighting a set of specific practices that may provide a starting point for measurement, recognizing the limits of a post-positivist approach and the potential benefits for connecting this conceptual material to other constructs such as organizational citizenship.

Professional civility is a framework for thinking about work, about the place where we work, about others with whom we work, and about the way we work with those others within a "local home." Professional civility involves respect for one's place of employment and those who work there, including oneself. Professional civility engages work and productivity within a background of respect for work, institution, and others.

Professional civility is an attitude manifested in behavior, or a way of engaging the world that translates to engaging work, one's organization, and other people. It is more than politeness, more than etiquette, more than "being nice," though it may bear the marks of these features. Professional civility has roots in dialogic civility (Arnett 2001; Arnett and Arneson 1999). Dialogic civility embraces hope about life together that is tempered by the knowledge that disappointment will meet us but that community connected through a common center of labor worth doing is ultimately necessary and worthwhile. In this manner, the world engaged with professional civility is not held to impossible standards, which would lead to cynicism generated from unmet high expectations.

Professional civility protects and promotes work and the institution: productivity. It recognizes the value of labor worthy of being done. Professional civility also recognizes the value of leisure and rest as well and the value of private life; each sphere of life is necessary for human beings to be healthy and whole. Professional civility does not mean burning oneself out or destroying oneself for work but doing the best one can while engaged in that work, honoring the place you work, and treating others who labor with you with consideration. Professional civility orients toward the larger good, the larger institution. Its primary focus of attention is not so much the other as much as a common project and/or organization; it knows that a focus on the institution will require considerate behavior toward others because that contributes to healthy institutions, and employees can be only as healthy as the institution. When we protect institutions, we provide spaces within which persons can flourish and find health.

Members of organizations who practice professional civility are oriented together toward a common project. Imagine this physical image, one of multiple people facing inward, aligned in a circle, with their arms upward. Together they hold up a structure. It is this joint operating together toward a common purpose that marks professional civility in its organizational and interpersonal dimensions.

Professional civility protects and promotes the good of coworkers. To be civil in the professional world, to practice professional civility, means to engage others with whom one works with consideration and respect, with a balance between the closeness that comes from shared work and appropriate distance—the type of distance that grants dignity and space to others, not the type that denotes coldness or aloofness. To practice professional civility is to engage in public discourse, to practice restraint in what we say and do, to grant others personal and institutional space. It is not being unfriendly or stiff. It is putting others at ease; it is being interpersonally trustworthy. Others can trust you not to embarrass them in front of others; they can trust you to take what they offer and put the best light on it. For example, if someone offers and idea at work that you do not find helpful, build on it instead of tearing it down. Imagine this scenario where a committee has met to plan a holiday party at the end of the year:

Bill:	Let's ask everyone to come dressed as a favorite character or representative from their own tradition. We can give a prize for the most elaborate costume!
Instead of saying:	"No one wants to go to all that trouble. That's too hard,"
Donna might reply:	"I like the idea of having people share a tradition. What if everyone brings in a favorite song on a CD, or maybe a holiday recipe?"

Professional civility means paying attention to the particular. In a world of diverse preferences and expectations, it is vital to pay attention carefully to the clues all around, to learn to read expectations of an organizational home and of one's colleagues. Professional civility means not assuming one's ideas are the only ideas; professionally civil discourse leaves room for other options. It does not attack the character of others. It disagrees without being disagreeable. Professional civility requires cultivating practical wisdom, knowing when to speak and when to be silent, and when one chooses to speak, how to speak wisely and well. Not every opinion needs to be voiced. Professional civility involves recognizing when to take others aside to express a concern rather than stating it in front of everyone. Professional civility seeks to promote constructive work together.

Omdahl (2006) offers an illustration of a professionally civil discourse style:

> Professional civility facilitates momentum toward accomplishing the mission by minimizing non-mission-focused demands. Imagine a colleague arriving at work telling a long, emotionally-draining tale about a personal relationship. Contrast it with another colleague who greets colleagues warmly, shares a short, uplifting story about an enjoyable experience, and gets on about the day's work. Needless to say, the second colleague has done a better job of moving self and others toward work. Later that day when support staff has failed to call the copier repair person, imagine that the first colleague throws a fit, publicly shaming the support staff. The second colleague stops by the desk of the support staff and offers a face-saving reminder. In doing so, it is highly likely that the second colleague has enabled resolution of the problem without compounding the problem with unnecessary emotional distress. (280)

She also reminds us to recognize that expectations for professional civility must never be a mask for oppression or marginalization of those who have historically been excluded from participation. There are professionally civil ways to bring about change, just as Martin Luther King, Jr. advocated nonviolent resistance. Publicly stated expectations for inclusion and participation can be brought to the fore as a resource for real change in the culture and virtue structure of an organization, moving beyond mere compliance with policies.

Professional civility seeks the contributions of all, giving ideas from multiple sources a chance to vie for implementation, though not all ideas can be implemented, and each idea must be held up to the light of the institution's mission. Professional civility works with others, whether those others are liked or not, and engages distance through roles in order to keep focus of attention on a common project rather on one's personal feelings about another.

The roles we inhabit in organizations grant us public dignity. Roles provide expectations and boundaries for behavior; one can learn role expectations and practice them well, which generates a source of genuine competence and excellence in doing a job and yields public evidence of strong performance that can be documented. Roles provide a basis for fair evaluation so necessary for a climate of professional civility in any organization.

When we treat others according to the expectations for their roles, we move the focus of attention away from the person to the role. Ironically, honoring a role will honor the person; engaging a role does not mean forgetting someone's humanity. Depersonalizing does not mean "dehumanizing." In certain situations, depersonalization can rehumanize. In a longitudinal study of turning points in problematic work relationships (Hess, Omdahl, and Fritz 2006), participants reported each week on a problematic relationship. One participant, a nurse, reported each week about her experiences with a very controlling boss, someone who simply drove her mad with her extreme structure, detail, and insistence on "bossing" everyone. Then, some weeks later, an entry appeared that ran something like this:

> "I now see my boss in an entirely new light. A family came in whose child had died. They were distraught, completely at a loss as to what to do. My boss took charge of the situation and brought order to their emotional chaos. Her structure was exactly what they needed. I now see how these characteristics that troubled me so have a place and are needed."

This nurse participant's focus of attention moved away from this boss's "trouble" for her, personally, and onto the boss's role, where, in this context, those behaviors were helpful. Depersonalization, but not dehumanization, had occurred.

Professional civility is a way of engaging others that looks first for the best in those around you. Emotional intelligence is part of professional civility, as is attributional graciousness. For example, what alternative explanations can you make for two people who are going to lunch and do not ask you to come along? You have a choice to focus attention on an explanation that attributes their behavior to deliberate exclusion of you as a person ("they don't like me" or "they don't want me to be part of their group") or to a different explanation, such as their need to finish a work project by a deadline—a role-related attribution (e.g., Omdahl 2006). Professional civility involves choices about interpretations of others' behavior, a good will of mind and heart. Professional civility is akin to being able to dance in such a way that even those who do not dance well will look good when dancing with you—and it is also about not kicking others in the shins when they are doing their best to dance with you.

Professional civility knows how to stop inappropriate discourse and gossip by redirecting conversation, speaking firmly but calmly for another perspective if someone who is not present is maligned or the victim of social undermining and knows how to refrain from laughing at jokes at others' expense without becoming a "wet rag" or "fussy." Professional civility recognizes that connection with others emerging as a byproduct of work can contribute to an enjoyable and pleasant work experience and a stronger organization.

Professional civility means avoiding the impulse to go on and on about oneself in conversation; it knows how to invite others to join in conversation by orienting them to the topic under discussion when they join a group. Professional civility works to avoid topics of known disagreement and controversy in order to focus attention on what others have in common: the work at hand. Professional civility refuses to generate an illusion of connection at work through a community of complaint and exercises leadership in engaging the unexpected and unwanted in a constructive manner.

For example, when you report back to your department about the work done in a subcommittee, instead of describing every little altercation or disagreement that happened in the group or how hard it was to get people to focus and concentrate, simply report the finished work of the group. Then, of course, find a mentor to whom you can speak in private to seek advice about how to get work done in such subcommittees in a more effective way.

Imagine that the administration of a university is constructing what will soon be an office suite and that fall semester is near, with completion still far off. A wise department chair might take people aside and say, "It will not help to complain about how the office isn't complete yet. Everyone can see that it's still being constructed; such focus on what is not yet complete will not be helpful. Instead, just do your job—teach and do research. Maintain a positive and upbeat demeanor. Students will look to you for cues about how to think about the circumstances. Your behavior needs to tell them that we are fortunate to be getting a new facility—and that you are moving on with your work in the midst of temporary inconvenience." Such advice reminds us that in a social world marked by events with ambiguous meaning, we have the power to enact for others a constructive understanding of the unexpected. We can frame events that we would not have chosen in ways that offer new, different, and better meaning. Our circumstances can present hopeful opportunities through our constructive, creative response. When many are engaged in this hopeful attitude, the entire community's strength contributes to a healthy sense of forward movement. We're not pretending the "problem" doesn't exist; we're reconfiguring what some would call a problem into a slight bump in the road that gives us a chance to test our shocks.

A number of specific practices encompass civility in the workplace and other professional settings. This summary list, derived from the material provided earlier, provides tentative starting principles for professionally civil behavior, ranging from verbal courtesy to respect for others in one's physical decorum, from which items can be generated to develop an operational definition to be used in further research:

1. Practice supportive behaviors. If you are a supervisor encouraging better work performance, focus on what the employee is doing well, highlight specific actions that the employee can take to improve, and provide appropriate comparison groups (for example, everyone who is new has a learning curve; you're showing good progress; here are ways to reach your full potential). This approach is helpful for coworkers, as well.
2. Avoid using jargon that only one or two people in a group understand. This type of behavior creates a "clique" of "in the know" folks and excludes others.
3. Find topics of conversation in which all can participate, even people with diverse interests and very little in common, Include newcomers in conversation by catching them up on the conversation so far, as appropriate.
4. Avoid asking potentially embarrassing or personal questions of others in public. "So, how big of a fine did you have to pay for that speeding ticket?" "Jim looked mad at you. What did you do this time?" "Are you still trying to find a date after Sally backed out on you?" "Donna just got promoted to account manager. Weren't you after that position?"
5. Perform personal grooming in private, not in public (e.g., avoid clipping your nails in public).
6. Avoid profanity. In public contexts, this minimal common ground for discourse promotes a focus on cooperative action.
7. Others should not be forced to overhear intimate details of your personal life that you've related to someone else in a cell phone conversation in public. Privacy protects both you and others.
8. Support the institution that provides your paycheck. If there are policies you disagree with, make those concerns known through formal channels in your organization. The practice of professional civility involves expressing public support for one's organization, even if one disagrees with some of its policies or practices. The metaphor of being a guest in someone else's home might work here for someone who does not feel part of an organization but is, nonetheless, employed by it or associated with it.
9. Apologize when you make a mistake, offend someone, or behave in an uncivil manner. We will step on each other's toes and need to extend interpersonal and professional grace to each other.
10. Practice forgiveness when others offend you.
11. Cultivate a rich private life to provide sources of enjoyment when circumstances are not as you hoped they would be at work.

Professional civility as an ethical communicative style in organizational settings requires mindfulness and attentiveness to become an alternative to communicative interpersonal crisis in the workplace. These reflections are a starting point for future research and scholarship.

REFERENCES

Andersson, Lynne M. and Christine M. Pearson. 1999. "Tit for Tat? The Spiraling Effect of Incivility in the Workplace." *Academy of Management Review* 24: 452–471.

Arnett, Ronald. C. 2006. "Professional Civility." *Problematic Relationships in the Workplace*, eds. Janie M. Harden Fritz and Becky Lynn Omdahl. NewYork: Peter Lang, 233–248.

Arnett, Ronald. C. 2001. "Dialogic Civility as Pragmatic Ethical Praxis: An Interpersonal Metaphor for the Public Domain." *Communication Theory* 11: 315–338.

Arnett, Ronald. C. and Janie M. Harden Fritz. 2001. "Communication and professional civility as a basic service course: Dialogic praxis between departments and situated in an academic home." *Basic Communication Course Annual* 13: 174–206.

———. 2003. "Sustaining Institutional Ethics and Integrity: Management in a Postmodern Moment." *Institutional Integrity in Health Care*, ed. A. S. Iltis. Dordrecht: Kluwer, 41–71.

Arnett, Ronald C., Janie M. Harden Fritz, and Leeanne M. Bell. 2009. *Communication Ethics Literacy: Dialogue and Difference*. Thousand Oaks, CA: Sage.

Baron, Robert A. and Joel H. Neuman. 1998. "Workplace Aggression—The Iceberg Beneath the Tip of Workplace Violence: Evidence of its Forms, Frequency, and Targets." *Public Administration Quarterly* 21: 446–464.

Barrett, Harold. 1991. *Rhetoric and Civility: Human Development, Narcissism, and the Good Audience*. Albany, NY: SUNY Press.

Becker, Howard S., and Irving Louis Horowitz. 1970. "The Culture of Civility." *Transaction* 7(6): 12–19.

Bennett, Rebecca J. and Sandra L. Robinson. 2000. "The development of a measure of workplace deviance." *Journal of Applied Psychology* 85: 349–360.

Brown, Penelope, and Stephen Levinson. 1978. "Universals in Language Use: Politeness Phenomena." *Questions and Politeness: Strategies in Social Interaction*, ed. Esther N. Goody. Cambridge: Cambridge Univ. Press, 56–310.

Bryant, Christopher G. A. 1993. "Social Self-organization, Civility and Sociology: A Comment on Kumar's 'Civil Society.'" *British Journal of Sociology* 44: 397–401.

Chapman, Joann. 1993. "Collegial support linked to reduction of job stress." *Nursing Management* 24(5): 52–56.

Collier, Mary Jane. 1994. "Cultural Identity and Intercultural Communication." *Intercultural communication: A reader,* ed. Larry A. Samovar and Richard E. Porter. 7th ed. Belmont, CA: Wadsworth Publishing, 36–45.

———. 2003. "Understanding cultural identities in intercultural communication: A ten-step inventory." *Intercultural communication: A reader,* ed. Larry A. Samovar and Richard E. Porter. 10th ed. Belmont, CA: Wadsworth Publishing, 412–429.

Coombs, W. Timothy and Sherry J. Holladay. 2004. "Understanding the Aggressive Workplace: Development of the Workplace Aggression Tolerance Questionnaire." *Communication Studies* 55: 481–497.

Gant, Larry M., Biren A. Nagda, Howard V. Brabson, Srinikia. Jayaratne, Wayne A. Chess, and Anup Singh. 1993. "Effects of social support and undermining on African American workers' perceptions of coworker and supervisor relationships and psychological well-being." *Social Worker* 38: 158–164.

Giacalone, Robert A., and Jerald Greenberg. 1997. *Antisocial Behavior in Organizations*. Thousand Oaks, CA: Sage.

Hess, Jon A., Becky Lynn Omdahl, and Janie M. Harden Fritz. 2006. "Turning Points in Relationships with Disliked Coworkers." *Problematic Relationships in the Workplace*, ed. Janie M. Harden Fritz and Becky Lynn Omdahl. New York: Peter Lang, 89–106.

Hochschild, Arlie R. 1983. *The Managed Heart: Commercialization of Human Feeling*. Berkeley: University of California Press.

Kimura, Heather. 2003. "Overcome toxic management." *Nursing Management* 34: 26–29.

Kumar, Krishan. 1993. "Civil Society: An Inquiry into the Usefulness of an Historical Term." *British Journal of Sociology* 44: 375–395.

Lutgen-Sandvik, Patricia. 2003. "The communicative cycle of employee emotional abuse." *Management Communication Quarterly* 14: 471–501.

Metts, Sandra, William. R. Cupach, and Lance. Lippert. 2006. "Forgiveness in the workplace." *Problematic Relationships in the Workplace*, ed. Janie M. Harden Fritz and Becky Lynn Omdahl. New York: Peter Lang, 249–278.

Namie, Gary and Ruth Namie. 2000. *The Bully at Work. What You Can Do to Stop the Hurt and Reclaim Your Dignity on the Job*. Naperville, Illinois: Sourcebooks.

Omdahl, Becky Lynn. 2006. "Towards effective work relationships." *Problematic Relationships in the Workplace*, ed. Janie M. Harden Fritz and Becky Lynn Omdahl. New York: Peter Lang, 279–294.

Omdahl, Becky Lynn and Janie M. Harden Fritz. 2006. "Effects of negative interpersonal relationships on stress and well being." *Problematic Relationships in the Workplace*, ed. Janie M. Harden Fritz and Becky Lynn Omdahl. New York: Peter Lang, 109–130.

Pearson, Christine M., Lynne M. Andersson, and Christine. L. Porath. 2000. "Assessing and Attacking Workplace Incivility." *Organizational Dynamics* 29: 123–137.

Pearson, Christine M. and Christine L. Porath. 2005. "On the nature, consequences, and remedies of workplace incivility: No time for 'nice'? Think again." *Academy of Management Executive* 19(1), 7.

Simons, Tony. 2002. "Behavioral integrity: The perceived alignment between managers' words and deeds as a research focus." *Organization Science* 13: 18–35.

Sullivan, William. 1995. *Work and Integrity: The Crisis and Promise of Professionalism in America*. New York: HarperCollins.

Sypher, Beverly D. 2004. "Reclaiming civil discourse in the workplace." *Southern Communication Journal* 69: 257–269.

Tehrani, Noreen, ed. 2001. *Building a Culture of Respect: Managing Bullying at Work*. New York: Taylor and Francis.

Vardi, Yoav, and Ely Weitz, eds. 2004. *Misbehavior in Organizations: Theory, Research, and Management*. Mahwah, NJ: Lawrence Erlbaum Associates.

Walzer, Michael. 1974. "Civility and Civic Virtue in Contemporary America." *Social Research* 41, 593–611.

Wood, James R., and Jon P. Bloch. 1995. "The Role of Church Assemblies in Building a Civil Society: The Case of the United Methodist General Conference's Debate on Homosexuality." *Sociology of Religion* 56: 121–136.

Zauderer, Donald G. 2002. "Workplace incivility and the management of human capital." *The Public Manager* 31: 36–42.

Questioning Back

Engaging an Organization's Narrative for Ethical Communicative Responsiveness in Crisis Situations

S. Alyssa Groom
Duquesne University

As with any crisis, this essay begins with a situation.

> Two Buddhist monks were returning from a pilgrimage. It was spring, and the rains had fallen. They came to a river, which was swollen and running swiftly so that the steppingstones were covered with water. A young girl stood by the river, afraid to cross. The senior monk simply picked her up, sloshed across, set her down on the other side, and continued on his way.
>
> About an hour later, the younger monk said, "Forgive me, I know you are older and wiser than I and have been longer in the religious life, but do you really think that it was right for you, a celibate monk, to pick up that young girl in your arms and carry her that way?" The older monk replied, "Oh my son, are you still carrying her?"

This story reveals two distinct ideas about the way in which communicative responses are formed in moments that demand ethical crisis responsiveness. The first idea emerges from the response of the older monk. His could be understood in terms of Heidegger's (1953, 1996) conception of being and its existence "within-time"—a "withinin-timeness" responsiveness. The older monk can be assumed to be both older and wiser. He knew his path well. He assessed the situation and responded within the moment to address the "crisis" of passage across the river for a young girl. Older and more experienced, he lived from a circumspect perspective. The older monk new what it meant to be a monk both in terms of his present embeddedness in the tradition and his responsibility to both the past and future of that tradition. It was this understanding that enabled him to respond to the young girl as he did, without violating his tradition. His existential response moved him confidently in the direction of "being-as-a-monk" living toward the other vis-à-vis an "existential concept of care" (Heidegger, 1953/1996, 186).

The older monk's continuous movement appeared to reduce the distance between his action toward the young girl and his call to lead a celibate life to a point of seamless proximity when, in fact, they were no closer than before he encountered the situation. Upon closer examination, it becomes clear that his act revealed a focus of attention established in a greater revelation of the path on which both he and the younger monk traveled. For the older monk, that path was part of the continuous movement of being towards Being—movement towards the fullness of a life constituted by his situatedness as a monk.

In contrast, the younger monk provided a response emergent from within a more familiar topography. His question to the older monk revealed a position of heightened self-awareness and what Charles Taylor (1989) termed a localized sense of moral sources in which "the localization is *ours* . . . from *within* which we experience and deliberate about our situation" (111). From his youthful standpoint, the younger monk experienced the situation, contemplated the action of the older monk, and projected an idea of what should have been done in the form of a question raised after the experience had come and gone. While the younger monk encountered the same situation as the older monk, he derived his perspective and response from his own source of understanding about the appearance associated with being a monk and not from a "living" or true sense of what it meant to walk through life as a monk.

The claim could be made that both the older and the younger monk "questioned back" to engage a response to the situation. The younger monk questioned back with adjunct definitiveness emergent from positional certainty—the ability to make a claim about exactness grounded in one's own understanding. His positional certainty focused his attention only on the appearance of life as a monk. His idea of what it meant to live as a monk kept him focused on the moment. The older monk, however, questioned back through what Gadamer (1975/1998) termed a historically effected consciousness. In a moment of crisis, he engaged, acted, and then walked on. In the wisdom and action of the older monk is revealed a fundamental awareness of "all human thought as historically conditioned" yet not derived solely from an unconditional standpoint (Gadamer 1975/1998, 448). The older monk's action and continuous movement illustrated this consciousness and an ethical responsiveness amidst the inherently contradictory nature of life.

It is the wisdom and action of the older monk that compels this interpretive investigation into ethical crisis communication. Writer Madeline L'Engle (2002) wrote about the origin of such wisdom as "the Rock that is higher." This rock is a shelter, anchor, protection, defense, plumb line, and stable orientation upon which a firm dwelling is established. Though she wrote of the rock in spiritual terms, L'Engle placed a demand in her explication on "story as truth"—a demand worth considering beyond the realm of the spiritual.

For organizations facing a crisis situation, questioning back within time is the existential engagement of a rock that is higher. Questioning back to the narrative of a given organization invites the "narrativity" of that organization to speak into a crisis situation as the rock that is higher, and to inform an ethical communicative response (Ricoeur 1983/1984). Questioning back takes seriously the narrative commitments of Paul Ricoeur (1983/1984), who said "narrative is in no way bound to the confused and limited perspective of the agents and the eyewitnesses of the events. On the contrary, the putting at a distance that constitutes a 'point of view' makes possible the passage from the narrator to the historian" (178). This dialectic engagement embraces the tension of appropriation and distanciation inherent in all decision-making, especially those characterized as a crisis. This tension is marked by the desire to make something one's own while simultaneously recognizing the inevitability of an estrangement that cannot be fully reconciled in the act of appropriation. Questioning back to the narrative of an organization during times of crisis attends to an ever-present dialectic and to a point of engagement that is "higher" than the current situation, adding a valuable dimension to the response.

The twofold hermeneutic entrance into this project unites crisis communication and narrative theory in an interpretive look at questioning back as a necessary act in moments of organizational crisis. Ian Mitroff's pioneering work on crisis management provides an applied starting point for engaging crisis communication. His strategic perspective embraces the ambiguity and complexity of true crisis while offering a method for addressing such situations. The art of questioning back as presented through Paul Ricoeur's narrative theory enlarges the framework of engagement surrounding moments of crisis and textures methodology with narrative insight. For Ricoeur, narrative simultaneously embraces the estrangement and proximity associated with the "text." Narrative permits a commitment to the past, present, and future to co-inform one another. In the instance of a crisis situation, a commitment to a co-informed perspective creates the possibility of gaining otherwise unattainable insights into the situation. Together, the interpretive union of crisis responsiveness and narrative on the subject of places narrative in the moment of crisis and invites an organization's forward movement based on the existential presence of a rock that is higher. In questioning back to its narrative, an organization lives as the older monk, engaging in responsiveness while walking.

THE IMPORTANCE OF QUESTIONING BACK

As media consumption of companies and people in crisis continues to escalate in the wake of significant personal, corporate, and government scandals, so too does the need for ethical crisis communication. By most standards, crisis measures in organizational contexts include a methodological approach that attempts to pre-plan and/or anticipate what questions to ask, a framework for engaging message content, a system of management for the communication that will occur within an organization and between an organization and its publics, and the forward movement of the organization beyond the crisis itself. Crisis situations, however, are not linear in their emergence or impact, making it difficult to enact a pre-planned response. Complexity theory suggests that they are ambiguous, unpredictable, uncertain, and unstable (Gilpen and Murphy 2008). Questioning back is one way of providing a complex crisis situation with a stable point of engagement from which to respond.

In this section, the importance of questioning back is examined in relation to an organization's commitment to a "community of memory" and to ongoing conversation in a given historical moment (Bellah, et al. 1985, 152; Arnett and Arneson 1999, 273). As crises emerge from within communicative contexts that, like a prism, are multifaceted, attentiveness to a community of memory and to the ongoing conversation remind an organization of its presentness in particular time and a particular place. They also connect it to both the past and future through ongoing engagement with the ideas that formed and continue to sustain the organization. The communicative nature of these ideas, particularly as they form an organization's narrative, enable a "web" of associations that help guide an organization in times of crisis (Geertz 1973; Taylor 1989; Arnett and Arneson, 1999).

QUESTIONING BACK TO A COMMUNITY OF MEMORY

Derived from Tocqueville's (1840/2002) exploration of America, an understanding of what it means to exist in a community of memory found renewed vitality through the work of Bellah, et al. (1985). Their sociological investigation into the associative and individualistic nature of American life publicly framed community in a sense of "historicality" and the "stories and dramas that bind the people (of that community) together" (Arnett and Arneson 1999, 272–3). Continued development of this idea through what Arnett and Arneson (1999) term "story-centered values" offers people a language and associative framework metaphorically reminiscent of Tocqueville's (1835/2002) observation that the bond of language was one of America's most enduring and defining characteristics (272). In the early years of the nation, language united people across beliefs, individual pursuits, culture, and heritage towards a common purpose. The connection to a community of memory works within the spirit of this idea, offering a given drama or story as an alternative language that binds people together through situated understanding, involvement, and participation in everyday life.

The story emergent from a community of memory affirms a set of goods or ends that stand independent of individual conceptions of the good as well as individual pursuits. Existence thus becomes a situated enterprise, formed and shaped within a "space of moral issues" dealing with "identity and how one ought to be" (Taylor 1989, 199). Being able to "find one's standpoint in this space (story), being able to occupy, to *be* a perspective in it" is inextricably linked to an understanding of one's situatedness amidst "webs of interlocution" (112, 36). Identity and stability are referential as relations with interlocutors—participants in particular stories—nurture and constitute an understanding of existence.

While considered extremely relevant to interpersonal and communal contexts, these ideas about a community of memory can also be applied to organizations through the broader notion of narrative. Ricoeur (1984), Fisher (1984), MacIntyre (1984), and Taylor (1989) are all seminal contributors to the conversation on narrative. All offer constructive insights into narrative theory and engagement. And while each differs in his starting point—from literary theory to theories associated with the communicative, interpersonal, and sociological aspects of life—they all converge on one issue: narrative understanding is at the center of the human condition.

Since the purpose of this paper is not to explicate narrative per se, what follows is an encapsulated review of ideas that embody this convergence. Ricoeur (1984) declares the storied nature of narrative as a means for understanding "both the language of 'doing something' and the cultural tradition from which proceeds the typology of plots" (57). Human action is narrated action and can only be understood because it is narrated through a relation to signs, rules, and norms that already exist (Ricoeur 1984, 57). Narrative is the means by which we grasp together the episodes of life creating a followable framework from which to engage and respond to contingencies, expectation, representation, and our own embeddedness (emplotment) (Ricoeur 1985). Fisher's (1985) narrative paradigm insists that discourse "is always more than the individuated forms that may compose it" (347). Narrative stresses ontology and is the means by which we order and understand the episodes of life. For Fisher (1987), narrative is established on the fact that all human communication is based in story; our ability to pull stories together in a coherent way depends on the human ability to achieve narrative rationality. MacIntyre's (1984) engagement of narrative positions it as the unifying factor in human life: "The unity of a human life is the unity of a narrative quest" (219). Narrative constitutes a "living tradition" by which the "phenomenon of embedding" permits us to examine the histories that surround and ground us and thereby make sense of the world in which we live. Lastly, Taylor (1989) regards narrative as means for establishing where we are and where we are going (47). Narrative is how we "grasp" and make sense of our lives (47). We are present in narrative.

Ricoeur, Fisher, MacIntyre, and Taylor announce the significance of narrative as a mode of human existence, making overt the necessity of narrative for existential coherence. In terms of an organization, this coherence is no less important. Organizational narratives have been conceived as story-like research on a given institution or product line, collections of organizational stories that tell about the organization, and research about organizational life as story making and reading (Czarniawska 1997). In her conclusion about narrating the organization, Barbara Czarniawska (1997) points to the work of Lejeune (1989) on autobiography and the potential for dialectical narrative, Bahktin (1928/1985) on dialogical narrative, and Silverman and Trode (1980) on interruptive narrative to illustrate that multiple approaches to the relationship between narrative and an organization exist (205). Her conclusion declares the inclination within each of these perspectives to be preservationist. In action, they work to uphold convention. As an alternative, she calls for something between realism and fiction: ergonographics, or "writing narratives on work organizations" (202). What is most relevant to this work is not the debate between narrative approaches but Czarniawska's overt connection between narrative as having a language and construction capacity within organizations. Narrative builds and constitutes meaning(s) in an organization. And while reality and fiction may, at times, be difficult to distinguish from one another, the importance of narrative's constitutive function remains. It becomes the language that unites people to an organization and its purpose, identity, and presence in the marketplace.

Like a community of memory forms the language that articulates the dramas and the stories of people's lives, narrative forms the language that articulates the unfolding and shaping of an organization. In that articulation, a philosophical (and physical) sense of place as well as purpose takes root. Narrative rootedness of an organization grasps together the organization's identity and surroundings, establishing a framework from which "intellectual honesty" and recognizable responsibility associated with public action might emerge (Weil 1949/1997, 195). Questioning back to this position of narrative rootedness in crisis communication situations makes visible the language of action fostered in the heritage or cultural tradition from which proceeds the crisis or plot that is currently unfolding (Ricoeur 1984, 57).

The invaluable role of questioning back is in the integration of potentially episodic moments in an organization's life that would otherwise be lost in the moment to the detriment of long-term organizational success. Narrative's resemblance to a community of memory elevates as the power of collective memory to a living tradition that has the power to sustain and grow an organization, especially in times of crisis. A return to organizational narrative—a questioning back—invites discernment through the narrative continuity between story and history as well as the current historical moment. Questioning back reminds us of a "field of dreams" experience in which the love of a sport—its heritage, lineage, and established practices—is made visible in the present by questioning back. In the context of an organization facing a crisis, it is questioning back that not only brings discernment but also continues to make the organizational narrative real, relevant, and responsive.

Questioning Back and Conversation

In organizational crisis situations, the community of memory formed in and through narrative is made present only in public discourse. Today, however, technology and media practices encourage and even advocate on behalf of a new public expectation of complete transparency in which immediate answers are privileged over careful conversation about how to resolve the crisis. Therefore, organizational narrative is not necessarily a privileged voice in the response. Immediacy has made it so that the hidden and mysterious domains of an organization's johari window are exposed, often with significant public and marketplace consequences (Mitroff 2001). This continued exposure reinforces the fact that if a company is not forthright with its publics, there will be public retribution. It is, therefore, imperative that the public conversation of an organization always works with an understanding that crises are certain to occur and must be addressed both internally and externally.

When discussing measures for managing crises before they happen, Mitroff (2001) offers methodological consideration and insight into what preparation for maintaining organizational integrity might look like. In one of his more recent books titled *Managing Crises Before They Happen: What Every Executive and Manager Needs to Know About Crisis Management*, he looks back to four historical figures as crisis advisors, hypothesizing their response to a specific crisis situation in order to make visible a spectrum of potential advice that could be offered in any crisis situation. This spectrum ranges from a minimalist ethical response to a response that maximizes opportunities for personal, spiritual, and social development. The spectrum illustrates the reality that crisis situations will evoke a variety of perspectives on how to respond, none of which are entirely correct and/or appropriate when considered independently of one another. His point is that what is most necessary is "the conversation and interplay between these different kinds of advisors [perspectives]"—a conversation that is not yet a well-established aspect of most organizational crisis management efforts today (Mitroff 2001, 79).

Mitroff is clear that no matter what advice may come from crisis advisors, truth-telling is vital to protecting organizational integrity; it is the essential core of all crisis communication strategies. However, he is also clear that the importance of communicating the truth cannot be done with integrity apart from the conversation between different perspectives. Vantage point, or the perspective on how to handle crisis communication and/or prepare in advance for engaging crisis situations, provides a necessary context for responsiveness. Conversation is the catalyst for different vantage points to illuminate and texture the crisis situation. It also takes into consideration the more recent examinations of complexity theory that question methodological approaches to crisis management and advocate the importance of working within the multiple influences that exert themselves on a given situation (Gilpel and Murphy 2008, 5).

The act of viewing a situation from multiple perspectives not only illuminates but also has the potential to reveal more than was initially perceived. With conversation between perspectives as an essential starting point, Mitroff looks back—questions back—to gain insight into what might be done. The questioning back is both historical and grounded in what Gadamer (1975/1998) calls a "historically effected consciousness." On the one hand, the questioning back calls forward historical figures—people from another time and place whose wisdom might illuminate the current situation. These historical figures respond as themselves, so their interpretations bear the imprint of their own historical moment; their response is representative of who they are as historical figures. On the other hand, the questioning back requests a situated response established in the crisis of the current historical moment. Like Gadamer's (1975/1998) engagement of a historically effected consciousness, these historical figures are transposed, finding themselves within the communicative demands of a "web of historical effects" (300). They are limited by their "range of vision" or "horizon;" yet this same horizon permits them the ability to "see beyond" (Gadamer 1975/1998, 302). Ultimately, the historically effected consciousness is represented by their ability to work in the situation "in order to understand it" (303).

Ricoeur adds to this understanding a more pronounced distinction between the past and the present. Conversation that has historical distance acknowledges the "nearness of the remote . . . the efficacy at a distance" in which the paradox of otherness becomes a critical vantage point (Ricoeur 1986/1991, 73). As such, the vantage point offered by narrative acts brings to the foreground a repository for conceiving symbolic relationships, not the least of which is how to "stand" as an organization in the midst of crisis. Narrative offers both an abstract and concrete presence, inviting a "community of interpretation" by which meaning might be transferred (Ritivoi 2006, 75).

The importance of questioning back in conversation to an organization's narrative is thus not in finding certitude but in exposing a horizon of inquiry from which meaning and guidance might be discerned. The overarching ethical imperative is the communicative necessity of a question back "to" an organization's narrative in order to gain insight into the conversation here and now. Once clearly stated in the conversation, the conversation assumes a purposeful dynamic in crisis responsiveness characterized by the dialectical engagement of past and present in anticipation of the future.

The role of conversation stretches Mitroff's methodological sensibilities into more contemporary complexity-based approaches to crisis management because it privileges multiple vantage points and voices instead of rational responsiveness. While not fully aligned with the complex adaptive processes of Gilpin and Murphy (2008) in which "uncertainty, adaptiveness, and improvisation replace certainty, goal orientation, and control," conversation and a questioning back do embrace the need for a middle ground in which an informed response is an adaptive one (177).

Questioning back in conversation privileges the organization's narrative as an essential voice in crisis responsiveness. Action back to the organization's roots acknowledges the limits of responsiveness as more than the imperative of immediacy. Questioning back intensifies the truly ethical imperative of attending to the community of memory in which the organization is established. History and historicality, reality and fiction are counterparts in this existence; as such, they must be reconciled in order to appropriate a timely response. Crisis responsiveness is thus a dialectical act that recognizes limits and in the act of questioning back appropriates solutions toward desired outcomes that respect the organization and its publics.

QUESTIONING BACK: ENACTING THE DIALECTIC OF APPROPRIATION AND DISTANCIATION

This essay has considered the act of questioning back to an organization's narrative as the reconciliation of past, present, and future in the act of decision-making during crisis situations. Through the language of an organization's narrative (community of memory) or through touch points established in conversation, questioning back enacts a commitment to the existential nature of the organization and its presence in a particular time and a particular place. It simultaneously calls the organization into response-ability for its history and rootedness within certain values. The emphasis on complete public transparency is recalibrated amidst the recognition of a rock that is higher—the organizational narrative—and attention is turned toward adapting in complex and often ambiguous or unexpected situations.

In this final section, questioning back is presented as an ethical enactment of the dialectic between appropriation and distanciation. Explored in the work of Ricoeur (1976), this dialectic is inherently connected to the act of questioning back and to narrative itself. Appropriation and distanciation function as dynamic counterparts to one another, facilitating meaningful action through greater understanding of tradition and the "reception of historically transmitted cultural heritages" (Ricoeur 1976, 44). Together, they make explicit a tension in narrative and in all textual encounters (organizational crisis is considered here as a text) and support the importance of questioning back as a means for addressing this tension.

Appropriation: The Call for Complete Understanding

In his explication of the problem of writing, Ricoeur (1976) makes the following statement.

> The problem of writing becomes a hermeneutical problem when it is referred to its complementary pole, which is reading. A new dialectic emerges, that of distanciation and appropriation. By appropriation I mean the counterpart of the semantic autonomy, which detached the text from its writer. To appropriate is to make "one's own" what was "alien." Because there is a general need for making our own what is foreign to us, there is a general problem of distanciation. Distance, then, is not simply a fact, a given, just the actual spatial and temporal gap between us and the appearance of the principle of a struggle between the otherness that transforms all spatial and temporal distance into cultural estrangement and the ownness by which all understanding aims at the extension of self-understanding. Distanciation is not a quantitative phenomenon; it is the dynamic counterpart of our need, our interest, and our effort to overcome cultural estrangement. Writing and reading take place in this cultural struggle. Reading is the *pharmakon*, the "remedy," by which the meaning of the text is "rescued" from the estrangement of distanciation and put in a new proximity, a proximity which suppresses and preserves the cultural distance and includes the otherness within the ownness (43).

Ricoeur's introduction to appropriation and distanciation situates the tension between the act of reading and writing. In the interplay of the two, there emerges a pull between estrangement and making something foreign one's own. Such tension produces opposition amidst an interweaving of the experience with the text. Textual independence is weighted by the encounter with a reader and the desire to appropriate the ideas found in the text for oneself.

Appropriation is, therefore, the act of taking something alien to oneself and attempting to reform it into something no longer strange. As discussed by Ricoeur, the pull of appropriation is toward a totalizing or what might be called a "complete" understanding of the text. In a Levinasian sense, appropriation reflects the totalization of otherness. It is accomplished in history, and "rests on the affirmation and the conviction that the chronological order of the history of the historians outlines the plot of being in itself" (Levinas 1961/1969, 55). In other words, what can be ascertained in the text of history is a full understanding of the nature of the text (or being). Such a claim violates the ethical condition of language—the preexisting "ethical plane"—in order to possess something that is wholly other (205).

In the act of questioning back during organizational crisis, the temptation to make something perfectly clear, transparent, or wholly one's own is ever present. Appropriation is manifest as a desire to attain complete understanding of the situation and all its component parts and to respond with flawless precision. Methodologies and predictable constructs are employed to overcome learning difficulties and ensure measurable outcomes. Appropriation becomes a means for actualizing "fast" meaning in the present moment of crisis. And though comprehension of this process of appropriation is obtained through interlocution, of which the constitutive parts can be present, as well as past, and future, appropriation reifies the interplay of history and historicality in order to execute accuracy.

Distanciation: Productive Estrangement

As a concurrent experience in the tension between the text and one's engagement of it, Ricoeur turns to the problem of distanciation and the unavoidable estrangement that occurs alongside appropriative measures. Distanciation, according to Ricoeur (1976), is the dynamic counterpart to appropriation's attempt to become "contemporaneous with past geniuses" (44). Productive distanciation permits one to engage the current historical moment from within a particular communicative framework or moment not its own.

In the act of questioning back, distanciation offers philosophical and practical insights into a situation vis-à-vis an alternative vantage point. Distantiation makes visible more than a temporal gap in the situation. The struggle between present situatedness, organizational narrative, and crisis coexist in this idea. It represents the constant struggle to overcome what is unknown and the inability to do so.

In both appropriation and distanciation, questioning back finds its perfect articulation. In the words of Ricoeur (1976), "the dialectic of distanciation and appropriation is the last word in the absence of absolute knowledge" (44).

In moments of crisis, especially as understood within complexity theories, this sentiment is widely shared. Through questioning back, the dual engagement of appropriation and distanciation offers the possibility of insight in the midst of competing approaches to responsiveness.

QUESTIONING BACK TO THE EXISTENTIAL ROCK THAT IS HIGHER

The ethical enactment of appropriation and productive distanciation is, at its core, the hope of questioning back. In the meeting of the two, interpretation—the work of questioning back—has as its end goal "an attempt to make estrangement and distanciation productive" (Ricoeur 1976, 44). In the context of crisis communication the dialectic of appropriation and productive distanciation represents this tension inherent in questioning back to an organization's narrative in order to move forward. It is a tension bound up in concordant discordance and the hope of harmony emergent from complexity and moments of apparent discontinuity.

Much like the experience of a tragedy, only without the guaranteed tragic ending, questioning back to the narrative of an organization during a crisis invites the heritage, mission, tradition, wisdom, stability, and coherence of the organization to speak into the instability of the current situation. While never able to eliminate estrangement or distanciation, the voice of an organization's narrative puts into relief the background from which a response must emerge. To this end, the productivity inherent in the ethical enactment of appropriation and distanciation in questioning back is tied to the existential encounter with a rock—the background—that is higher than the moment of crisis. It is higher not by physical position but by its ability to engender a necessary phenomenological focus of attention. From a vantage point still within the communicative context of the crisis, an organization's narrative serves as phenomenological point of orientation whereby action can be procured amidst uncertainty.

Questioning back to the narrative of an organization confronts us with the phenomenological perspective and wisdom of the older monk. In questioning back, a path is evident. In questioning back, there is a past, present, and future to consider. In questioning back, there is an established organizational framework for engagement. In questioning back, narrative is announced as a guiding voice. In questioning back, complexity and methodology are interwoven, offering the possibility of a steady, grounded, and forward moving organization in spite of a crisis situation. And in questioning back, organizations like the older monk, might find that they are able to walk through crisis and not be moved.

REFERENCES

Arnett, Ronald C. and Pat Arenson. 1999. *Dialogic Civility in a Cynical Age: Community, Hope, and Interpersonal Relationships.* Albany, NY: State University of New York Press.

Bakhtin, Mikhail Mikhailovich and Pavel Nikolaevich Medvedev. 1978. *The Formal Method in Literary Scholarship: A Critical Introduction to Sociological Poetics.* Trans. Albert J. Wehrle. Baltimore, MD: Johns Hopkins University Press. (Orig. pub. 1928)

Bellah, Robert. N., Richard Madsen, William M. Sullivan, Ann Swidler, and Steven M. Tipton. 1987. *Habits of the Heart: Individualism and Commitment in American Life.* New York: Perennial Library.

Czarniawska, Barbara. 1997. *Narrating the Organization: Dramas of Institutional Identity.* Chicago: University of Chicago Press.

de Tocqueville, Alexis. 2002. *Democracy in America.* Trans. Harvey. C. Mansfield and Delba Winthrop. Chicago: University of Chicago Press. (Orig. pub. 1835 and 1840).

Fisher, Walter R. "Narrative as a Human Communication Paradigm: The Case of Public Moral Argument." *Communication Monographs* 51 (1984): 1–22.

——. 1985. "The Narrative Paradigm: An Elaboration." *Communication Monographs* 52: 347–367.

——. 1987. *Human Communication as Narration: Toward a Philosophy of Reason, Value, and Action.* Columbia, SC: University of South Carolina Press.

Gadamer, Hans Georg. 1998. *Truth and Method.* Rev. Trans. Joel Weinsheimer and Donald G. Marshall. New York: Continuum. (Orig. pub. 1975)

Geertz, Clifford. 1973. *The Interpretation of Cultures.* New York: Basic Books.

Gilpen, Dawn R. and Priscilla J. Murphy. 2008. *Crisis Management in a Complex World.* Oxford: Oxford University Press.

Heidegger, Martin. 1996. *Being and Time.* Trans. Joan Stambaugh. Albany: State University of New York Press. (Orig. pub. 1953).

Lejeune, Philippe. 1989. *On Autobiography.* Minneapolis: University of Minnesota Press.

L'Engle, Madeline. 2002. *The Rock That is Higher: Story as Truth.* Colorado Springs, CO: Waterbrook Press.

Levinas, Emmanuel. 1969. *Totality and Infinity: An Essay on Interiority.* Trans. Alphonso Lingis. Pittsburgh, PA: Duquesne University Press. (Orig. pub. 1961).

MacIntyre, Alasdair. 1984. *After Virtue.* Notre Dame: University of Notre Dame Press.

Mitroff, Ian. 2001. *Managing Crises Before They Happen: What Every Executive and Manager Needs to Know About Crisis Management.* New York: Amacom.

Ricoeur, Paul. 1991. *From Text to Action: Essays in Hermeneutics.* Trans. Kathleen Blamey and John B. Thompson. Evanston, Ill.: Northwestern University Press. (Orig. pub. 1986)

——. 1976. *Interpretation Theory: Discourse and the Surplus of Meaning.* Fort Worth, TX: Texas Christian University Press.

——. 1984. *Time and Narrative, Volume I*. Trans. Kathleen McLaughlin and David Pellauer. Chicago: University of Chicago Press. (Orig. pub.1983)

——. 1985. *Time and Narrative, Volume 2*. Trans. Kathleen McLaughlin and David Pellauer. Chicago: University of Chicago Press. (Orig. pub. 1984)

Ritivoi, Andreea Deciu. 2006. *Paul Ricoeur: Tradition and Innovation in Rhetorical Theory*. New York: State University of New York Press.

Silverman, David and Brian Torode. 1980. *The Material World: Some Theories of Language and its limits*. London: Routledge and Kegan Paul.

Taylor, Charles. 1989. *Sources of the Self: The Making of the Modern Identity*. Cambridge: Harvard University Press.

Weil, Simone. 1997. *The Need for Roots: Prelude to a Declaration of Duties Towards Mankind*. Trans. A. F. Willis. New York: Routledge. (Orig. pub. 1949)

The Ethical Imperative of Significant Choice

Addressing Learning Styles in Crisis Messages

Timothy L. Sellnow
Deanna D. Sellnow
University of Kentucky
Steven Venette
University of Southern Mississippi

Following the 2001 attacks on the World Trade Center and the Pentagon, the persistent threat of terrorism on American soil became the "new normal" for United States citizens (Sellnow, Seeger, and Ulmer 2005). To combat this ongoing threat of a terrorist attack, the Department of Homeland Security, itself founded after 9/11, sponsors six centers of excellence situated in the academic community. Relevant to this study is the National Center for Food Protection and Defense (NCFPD). The mission of the NCFPD is to advance the security and safety of the nation's food supply through research, education, and outreach. The national center, housed at the University of Minnesota, is composed of "researchers from 19 other universities plus personnel from independent research facilities, state health and agriculture agencies, and professional organizations; private sector consultants; and industry advisors from agriculture and food industry companies" (National Center for Food Protection and Defense, n.d.). The authors of this paper are affiliated with the NCFPD.

One of the primary responsibilities of the NCFPD is to develop a comprehensive risk and crisis communication strategy for responding to terrorist attacks on the nation's food supply. The relevance of communication to counterterrorist activities is articulated clearly by O'Hair and Heath (2005):

"Communication serves as the lens through which responses to terrorisms are identified and strategized, and communication becomes the primary tool for counterterrorism activities" (4). The goal of the NCFPD risk communication group is to develop "best practices for active engagement of multiple audiences in effective risk communications prior to, during and after potentially catastrophic food bioterrorism incidents" (National Center for Food Protection and Defense, n.d.).

At present, the NCFPD has achieved its goal of developing a series of best practices for effective risk and crisis communication. The current challenge is to test these best practices and to translate them into pragmatic message strategies that can effectively reach the multiple audiences identified by the center. This paper provides a preliminary description of the NCFPD communication group's efforts to test the effectiveness of the best practices with multiple audiences and to develop the most effective and ethical means for communicating risk and crisis messages to a diverse public.

Essentially, this essay describes an exploratory study sponsored by the NCFPD that is designed to assess the degree to which risk and crisis messages are best communicated to multiple audiences. The contextual framework for the analysis is Nilsen's (1974) ethic of significant choice. The theoretical lens of this study is adapted from the perspective of experiential learning theory, as described by Kolb (1971, 1984). More specifically, we ground our needs assessment and message testing in Kolb's conceptualization of learning styles and the learning cycle model. Specifically, we assess the capacity for messages rounding the cycle of learning to satisfy the standards of significant choice. In the following pages, we, first, offer a rationale for using learning styles as an indicator of audience diversity and review the learning style literature. In doing so, we summarize the concept of significant choice and previous research in learning styles. Second, we provide some preliminary findings to our research. Finally, we summarize our research to date and speculate about future applications.

LEARNING STYLES AND CRISIS COMMUNICATION

Monumental crises such as Three Mile Island, Bhopal, and 9/11 underscore the importance of audience-centered risk and crisis communication. Yet, communicating in a manner that meets audience needs is complicated by the fact that audiences are diverse. For example, Kreps et al. (2005) found that "one of the biggest challenges to effective biodefense communication is making relevant information accessible and understandable to highly varied subgroups in society" (196). They identify "crafting effective messages for diverse audiences" as one of the "primary principles" for effective biodefense communication (192).

One global means by which to analyze audiences in crisis situations is to show empathy. James Lee Witt, director of the Federal Emergency Management Agency under the Clinton administration, and his coauthor James Morgan argue that empathy is "one of the most important roles for a leader helping others respond to crisis" (Witt and Morgan 2002, 147). Seeger, Sellnow, and Ulmer agree (2003) when they contend that the ethics of humanism and care should prevail in crisis communication. To do so, organizations and agencies must assess the needs of their publics and share information accordingly. Similarly, Borda and Mackey-Kallis (2004) maintain that "by tapping into what the public is thinking and feeling, a company can avoid basing their actions solely on what the media is saying, and concentrate on what people really perceive as the problem" (127). An empathetic communication style can account for the varying needs of complex audiences. For example, Becker and Thompson (2005) found that, "compared to those with a problem-solving active coping style, individuals with a ruminative, passive coping style are at greater risk for deteriorated health" (61). Effective risk and crisis messages, then, ought to address these varying emotional and physical needs of a diverse audience.

The temptation is for organizations and government agencies to establish one crisis management plan and then rely solely on that plan as the crisis evolves. Doing so, however, can lead to communication patterns that are insensitive to varying audience needs. For example, Marra (2004) argues that, although crisis management plans are important in managing crises, "their value appears to be overrated" (312). Rather than simply acting on a series of checklists, Marra encourages organizations to operate from a communication culture that recognizes and addresses the complex needs of their publics.

These audience needs should also be considered in terms of the larger narrative surrounding the crisis situation. Heath, McKinney, and Palenchar (2005) observe that "cooperative and competing narratives" surrounding terrorist events have the potential for "bringing people together and putting them at odds with one another" regardless of whether the narrative content is "scientific or mythical" (149). Thus, the messages surrounding a crisis or high-risk situation, such as terrorism, must consider the uncertainty and vulnerability that confound the situation. Ideally, risk communication invites the public to participate as an active partner in the risk mitigation process. In fact, Heath et al. (2005) suggest risk communication is likely to fail if it does not "increase residents' control in the decision-making process" (159).

THE ETHIC OF SIGNIFICANT CHOICE

For residents to experience control in managing threats of terrorism, we contend they must have adequate information. Nilsen (1974) explains the adequacy of information related to key decision-making contexts from an ethical perspective. Nilsen introduced the concept of significant choice to assess the ethics of persuasive messages. Paramount to Nilsen's work is the claim that a good share of human dignity resides in the capacity to make rational decisions. As Rowan (1995) argues, risk communicators often engage in persuasion to influence those decisions. Exposing listeners to unclear or biased information can sway the decision-making process in directions that are not in the best interest of the listener. Through the ethic of significant choice, Nilsen characterizes the ideal circumstances for objective and well-informed decision-making. Nilsen explains that listeners are empowered by the ethic of significant choice when the following five standards are met:

- Stakeholders are free from physical or mental coercion.
- The choice is made based on all available information.
- All reasonable alternatives are included in the discussion.
- Both short-term and long-term consequences are disclosed and discussed.
- Both senders and receivers of messages are open about any personal motives they have that may influence their decision-making.

These five standards provide a means of assessing the degree to which a risk communication message avoids bias and achieves the ethic of significant choice. Ultimately, dedication to the ethic of significant choice results in a marketplace of ideas where diverse points of view are shared, understood, and debated. This marketplace of ideas creates the synergy of information that is essential for significant choice.

While emphasizing the five standards for open communication, Nilsen (1974) cautions against several forms of communication that could diminish the opportunity for significant choice. Nilsen labels these forms of problematic communication "misinformation" (71). Misinformation can come in any of the following forms:

- Incomplete information
- Biased information
- Incomplete or inadequately defined statistical units
- Vague or ambiguous terminology
- Implied relationships between issues where no relationship exists (for example guilt by association)

- False sense of urgency or importance about the issue
- Highly emotionalized language that distorts meaning (71–72)

For significant choice to occur, then, communicators must both meet the five standards discussed earlier and avoid these seven forms of misinformation. It follows that as risk and crisis communicators, we should strive to meet these standards and avoid these pitfalls, as well.

The NCFPD advocates a communication strategy that empowers stakeholders, in a manner befitting of the ethic of significant choice, to respond appropriately during crisis situations. With this philosophy and these best practices in mind, a segment of the NCFPD communication team at the University of Kentucky proposes exploring the possible heuristic benefits of testing messages that address multiple learning styles. We contend that robust messages that address various learning styles are better able to provide balanced information that addresses the available alternatives, including short-term and long-term outcomes. Thus, shaping messages to round the cycle of learning is one means by which risk and crisis communicators can meet the standards of significant choice and avoid the pitfalls of misinformation. The next section provides a detailed rationale for creating and testing risk messages that round Kolb's (1984) four-stage cycle of learning.

LEARNING STYLE THEORY

For many academics, the mention of learning styles and learning style theory conjures more negative than positive connotations. A number of identifiable circumstances have given rise to this unfortunate reaction. Two of the most noteworthy reasons are (1) the plethora of disparate learning style models that each propose different terminology and (2) the inappropriate application of learning style research to stereotype learners. Consequently, this section offers an historical account of learning style theory as a means by which to address these concerns and to make a case for grounding our work in experiential learning theory and the learning cycle model as conceptualized by David Kolb (1984) for audience analysis and message formation regarding risk and crisis communication.

One major strike against learning style theories is rooted in the fact that so many models exist, each of which uses different terminology. In other words, no one universally accepted theory and set of terms exist. To clarify, a good number of scholars have published models for understanding the role preferred learning styles play in increasing learner comprehension, retention, and application of theoretical concepts. For example, Witkin and his colleagues (1962, 1977, 1978) conceptualized the notion of cognitive styles as they relate to the learning style preferences of individuals. They conclude that effective learning experiences must address the cognitive styles of both field dependent (FD) and field independent (FI) learners. Canfield and Lafferty (1974) posit that certain affective dimensions influence individuals to prefer one of six different learning styles or combinations of them. Similarly, Dunn and Dunn (1978) present five dimensions that mark various learning style preferences. Although these and other learning styles models differ in certain ways, they all contend that traditional, one-way, lecture-dominated message dissemination simultaneously privileges some individuals and places at-risk for failure other individuals. This conclusion alone makes learning style theory valuable to the purposes of message formation and assessment related to risk and crisis communication.

The other primary flaw in many learning style theories concerns their tendency to stereotype learners by placing them into categories and then assuming they learn best in one particular way. In other words, those who are labeled as visual learners or tactile learners or aural learners are characterized as able to comprehend, retain, and apply material best in ways that address only that one preferred style. David Kolb's (1984) model of experiential learning, however, overcomes this weakness by professing that, regardless of one's preferred learning style, true learning embodies comprehension, retention, and application and "involves the integrated functioning of the total organism—thinking, feeling, perceiving, and behaving" (31). Hence, Kolb purports that effective learning rounds an entire cycle that addresses each of four preferred learning styles. Although individuals tend to possess a learning style preference, all people learn best when given an opportunity to round the entire learning cycle.

Kolb's (1984) theory of experiential learning and learning cycle model are actually grounded in Dewey's (1938) "philosophy of education based upon a philosophy of experience" (29) that claims effective learning must somehow connect knowing with doing. For Kolb, effective learning occurs by rounding the learning cycle comprised of concrete experience (feeling), reflective observation (watching), abstract conceptualization (thinking), and active experimentation (doing).

Each of the four learning style preferences adheres to one of four stages of learning defined by Kolb. Learners who have a stage-one preference prefer a combination of concrete experience and reflective observation (divergers). That is, they tend to be motivated by lots of concrete examples and multiple perspectives. Learners who have a stage-two preference prefer a combination of reflective observation and abstract conceptualization (assimilators). These learners tend to prefer logically ordered facts, definitions, and explanations. Learners who have a stage-three preference prefer a combination of abstract conceptualization and active experimentation (convergers). That is, they tend to prefer solving real-world problems through practical applications. Learners who have a stage-four preference prefer a combination of active experimentation and concrete experience (accommodators). These learners tend to enjoy creative problem-solving strategies, teamwork, and thinking outside the box. No one preference is better than the others. In fact, the natural learning cycle is one where we move through all of the four stages (or learning style preferences).

Based on Kolb's (1984) theory, we argue that knowing the learning style preference of target audience members is important not to pigeonhole messages in ways that focus only on one style but, rather, to determine the optimum point of entry for rounding the learning cycle. We contend that point of entry may be important because starting with the stage that encompasses the primary learning style preference of audience members might initially motivate people to listen because doing so demonstrates empathy (Witt and Morgan 2002) and taps into "what the public is thinking and feeling" (Borda and Mackey-Kallis 2004, 127). Regardless of entry point, however, is the necessary goal to round the entire four-stage cycle (Kolb 1984).

We argue the rounding of the learning cycle can be operationalized in risk and crisis messages by formulating messages that address not only traditional information dissemination, but also discussions and activities. Doing so helps learners internalize ideas by conceptualizing a rationale for learning the material, becoming familiar with key concepts, envisioning a breadth of examples, and actively applying skills to real-life problems.

In this study, we collected descriptive data about learning style preferences in a general population and determined potential correlations to the best practices for effective risk and crisis communication.

METHOD

Participants

Participants (N = 249) were recruited by graduate students at a medium sized research university located in the Midwest. A snowball sampling technique was applied to reach a variety of individuals who were outside the university.

Procedures

Graduate students mailed the survey instrument to friends and family. For each survey mailed, the graduate students were paid a small honorarium. Graduate students were instructed to send the survey to people outside the region. This

Table 1. National Center for Food Protections and Defense 10 Best Practices in Risk and Crisis Communication

1. Communication specialists should participate in the decision-making process and in policy development when responding to a terrorist threat.
2. Pre-event planning is essential to effective risk and crisis communication. Plans should be updated constantly based on experience and new information.
3. Agencies should create a dialogue with and form partnerships with the public in planning for and managing terrorist events.
4. Agency leaders should listen to the public's concerns, regardless of perceived accuracy, and respond to them.
5. Agency spokespersons should communicate with honest, candor, and openness.
6. Agencies should make strategic plans to collaborate and coordinate with credible sources (e.g. subject matter experts, first responders) prior to a terrorist event.
7. Agency spokespersons should, to the best of their ability, meet the needs of the media and remain accessible.
8. Agency spokespersons should communicate with compassion, concern, and empathy.
9. Agencies should accept uncertainty and ambiguity as an inherent part of terrorism and avoid over-reassuring the public.
10. Agencies should provide some level of self-efficacy to the public.

request was made in hopes of collecting a data set from a broad geographic area. Respondents returned their survey instruments through standard mail. Confidentiality for all respondents was maintained. The university's internal review board approved the survey instrument.

Survey Instrument

The survey instrument was designed to assess crisis communication from several angles. The section of the survey relevant to this study sought to determine the extent to which the best practices established by the NCFPD (see Table 1) correlate with the four preferred learning styles described above.

PRELIMINARY RESULTS

As we stated at the outset, this project is currently in the exploratory phase. Our overall objective is to assess the degree to which message preferences vary among multiple audiences. Our first step was to determine the range of learning styles in our sample. Table 2 reveals the distribution of our sample among the four preferred learning styles.

Table 2. Distribution of Learning Styles

Learning Style	*Frequency*	*Percentage*
Divergers	113	45.4
Assimilators	32	12.9
Convergers	34	13.7
Accommodators	70	28.1
Total	249	100

Our second step toward completing this objective was to determine whether or not preferences for the best practices correlated with the preferred learning styles of respondents. To date, we have identified two best practices and several demographic characteristics where the variance among the learning styles is significantly correlated. The best practices with significant correlations are self efficacy (r [248] = –.186, p = .005): Agencies should provide some level of self-efficacy to the public; and updating plans (r [248] = .191, p = .004): Pre-event planning is essential to effective risk and crisis communication. Plans should be updated constantly based on experience and new information.

Self-Efficacy

For self-efficacy, participants responded to the following statement on a Likert scale: "During a crisis, I want to know what the government will do to fix the problem rather than being told what I can do to protect myself and my loved ones." Divergers and assimilators tended to agree with this statement, while convergers and accommodators tended to disagree. As such, divergers and assimilators wanted government agencies to address the problem and to protect them from any threats. Convergers and accommodators were more likely to desire information about what they could do to protect themselves during the crisis.

Updating Plans

The second best practice that showed a significant correlation with learning styles involved the best practice number two, updating plans. "The government should carefully assess its responses to crises so that new policies or laws can be based on what is learned." Overall, divergers and assimilators are less concerned about updated plans than convergers and accommodators. This finding suggests that divergers and assimilators may be less likely to embrace message modifications than convergers or accommodators. This pattern was also evident in a related question from the survey: "After a crisis, the government should try to forget what happened. Leaders should not dwell on mistakes that were made." Divergers and assimilators tended to agree with this statement, while convergers and accommodators disagreed. (r [248] = –.223, p = .001) Thus, divergers and assimilators may be less likely to respond favorably to post-crisis messages related to fault-finding and organizational learning than convergers or accommodators.

DISCUSSION

Based on these preliminary findings, a few patterns of message preference based on preferred learning styles are apparent. Messages of self-efficacy, for example, appear to be important to some members of the public and rather unimportant to others. To assure that participants perceive themselves as being understood (Witt and Morgan 2002) crisis communicators must develop messages that attend simultaneously to these varying preferences. Doing so honors the ethic of significant choice by addressing the multiple stages of the learning cycle. In other words, messages of self-efficacy also reassure the audience that the government and other appropriate agencies are doing all they can to protect the public and to resolve the problem.

These results also suggest that messages of organizational learning and crisis plan revisions resonate well with convergers and accommodators. Hearing about them is likely to stimulate these individuals who tend to enjoy active involvement in problem solving (Kolb 1984). Conversely, divergers and assimilators may be uncomfortable hearing messages about learning from failure and updating crisis management plans because doing so calls into question the security of a logical plan (1984). This, in turn, may heighten a perception of incompetent leadership, which may essentially provoke fear. To avoid such a reaction among divergers and assimilators, messages about organizational learning and about crisis plan updates should be communicated in a manner that reassures stability and order, and promises progress. For example, a message about a plan update could be communicated in a manner emphasizing that such updates are actually a normal part of the ongoing planning process. In doing so, divergers and assimilators might be reassured that the overall plan remains stable in that updates are essentially part of the preconceived plan.

The ultimate goal, then, is to develop concise but complete crisis messages that round the entire cycle of learning. This snowball sample assessment suggests further that the optimal point of entry on the cycle may, in fact, be stage one, since the largest number of respondents fell into that learning style preference (divergers). Hence, the messages ought to begin by offering a variety of examples both from previous research and from the multiple perspectives of the participants. Doing so might motivate participants in that it invites them to be "active participants in the . . . process" (Heath et al. 2005, 159). From there, the other stages (i.e., information dissemination, including key terms and concepts, practical applications in real life settings, and creative problem-solving) ought to follow.

CONCLUSIONS

The preliminary findings of this study suggest that the public varies in its message preferences. Crisis communicators must understand this variance if they are to gain compliance from the public. Kolb's (1984) notions of preferred learning styles and the learning cycle appear to be a useful means for understanding diverse preferences of various publics. Moreover, the four-stage model may be an effective tool for shaping messages that address the entire learning cycle, thereby honoring the standards of the ethic of significant choice. This preliminary study suggests that crisis messages can be designed to address a variety of learning style preferences. Specifically, messages of self-efficacy and planning updates can be constructed in a manner that simultaneously reassures stability and government protection, while emphasizing progress and self-protection.

Tailoring risk and crisis communication messages to round the learning cycle would ensure that the learning style preferences of diverse audiences are addressed, thereby making them "accessible and understandable to highly varied subgroups in society" (Kreps et al. 2005, 196). Moreover, doing so could potentially increase people's comprehension, retention, and application of best practices in risk and crisis communication (Kolb 1984). Further, shaping best-practice risk and crisis communication messages to appropriately address potentially counterproductive perceptions rooted in learning style preferences could improve compliance gaining among diverse publics (Seeger, Sellnow, and Ulmer 2003). Finally, understanding the primary learning style preferences of various audiences may help determine the most effective learning cycle point of entry that will tap into "what people really perceive as the problem" (Borda and Mackey-Kallis 2004, 127). Doing so is likely to heighten the motivation to learn, increase confidence in the message content, and ultimately to take appropriate action (Kolb 1984).

The results of this preliminary study suggest that Kolb's (1984) notions about preferred learning styles and the learning cycle may be useful for creating and assessing effective risk and crisis communication messages that honor the standards for the ethic of significant choice proposed by Nilsen (1974). Although the learning cycle approach is not the only means by which to do so, these studies suggest it is worth pursuing as one potentially valuable method.

REFERENCES

Becker, Jennifer A. and Sharlene Thompson. 2005. "The Intersection of Terrorism, Interpersonal Communication, and Health." In James A. Johnson, Gerald R. Ledlow, and Mark A. Cwiek (Series Ed.) and Dan O'Hair, Robert L. Heath, and Gerald R. Ledlow (Vol. Eds.), *Community Preparedness and Response to Terrorism: Vol. 3 Communication and the Media* (47–64). Westport, CT: Praeger.

Borda, Jennifer L. and S. Mackey-Kallis. 2004. "A Model for Crisis Management." In Dan P.Millar, and Robert L. Heath (Eds.), *Responding to Crisis: A Rhetorical Approach to Crisis Communication* (117–138). Mahwah, NJ: Lawrence Erlbaum Associates.

Canfield, A. A. and J.C. Lafferty. 1974. *Learning Styles Inventory*. Birmingham, MI: Humanics Media.

Dewey, John. 1938. *Experience and Education*. New York: Macmillan Publishing Company.

Dunn, Rita and K. Dunn. 1978. *Teaching Students Through Their Individual Learning Styles: A practical approach*. Reston, VA: Reston Publishing.

Heath, Robert L., David B. McKinney, Michael F. Palenchar. 2005. "Community right-to-know vs. terrorists' exploitation of public information." In James A. Johnson, Gerald R. Ledlow, and Mark A. Cwiek (Series Ed.) and Dan O'Hair, Robert L. Heath, and Gerald R. Ledlow (Vol. Eds.), *Community Preparedness and Response to Terrorism: Vol. 3 Communication and the Media* (125–166). Westport, CT: Praeger.

O'Hair, Dan H. and Robert L. Heath. (2005). "Conceptualizing Communication and Terrorism." In James A. Johnson, Gerald R. Ledlow, and Mark A. Cwiek (Series Ed.) and Dan O'Hair, Robert L. Heath, and Gerald R. Ledlow (Vol. Eds.), *Community Preparedness and Response to Terrorism: Vol. 3 Communication and the Media* (1–12). Westport, CT: Praeger.

———. 1971. Individual Learning Styles and the Learning Process. (Working paper No. 53571). Cambridge, MA: Sloan School of Management, MIT.

Kolb, David A. 1984. *Experiential Learning: Experience as the Source of Learning and Development*. Englewood Cliffs, NJ: Prentice-Hall, Inc.

Kreps, Gary L., Ken Alibek, C. Bailey, Linda Neuhauser, Katherine E. Rowan, Lisa Sparks. 2005. "The critical role of communication to prepare for biological threats: Prevention, mobilization, and response." In James A. Johnson, Gerald R. Ledlow, and Mark A. Cwiek (Series Ed.) and Dan O'Hair, Robert L. Heath, and Gerald R. Ledlow (Vol. Eds.), *Community Preparedness and Response to Terrorism: Vol. 3 Communication and the Media* (191–210). Westport, CT: Praeger.

National Center for Food Protection and Defense. (n.d.). www.fpd.umn.edu/index.html (accessed May 28, 2006).

Nilsen, Thomas R. 1974. *Ethics of Speech Communication* (2nd ed.). Indianapolis: Bobbs Merrill.

Rowan, Katherine. E. 1995. "What Risk Communicators Need to Know: An Agenda for Research." In Brant R. Burleson (Ed.), *Communication Yearbook Volume 18* (300–319). Thousand Oaks, CA: Sage.

Sellnow, Tim L., Matthew W. Seeger and Robert R. Ulmer. 2005. "Constructing the 'new normal' through post-crisis discourse." In James A. Johnson, Gerald R. Ledlow, and Mark A. Cwiek (Series Ed.) and Dan O'Hair, Robert L. Heath, and Gerald R. Ledlow (Vol. Eds.), *Community Preparedness and Response to Terrorism: Vol. 3 Communication and the media* (167–190). Westport, CT: Praeger.

Teen Workers Face Dismal Summer Job Outlook. 2004. NewsHour Extra. June 21. www.pbs.org/newshour/extra/features/jan-june04/teenjobs_6–21.html (accessed June 2, 2006).

Witkin, Herman A. 1978. *Cognitive Styles in Personal and Cultural adaptation.* Worchester, MA: Clark University Press.

Witkin, Herman A., C. A. Moore, D. R. Goodenough, and P. W. Cox. 1977. "Field-dependent and field-independent cognitive styles and their educational implications." *Review of Educational Research* 47: 1–64.

Witkin, Herman A., R. B. Dyk, H. F. Faterson, D. R. Goodenough, and S. A. Karp. 1962. *Psychological Differentiation.* New York: Wiley.

Witt, James Lee and James Morgan. 2002. *Stronger in Broken Places: Nine Lessons for Turning Crisis into Triumph.* New York: Time Books.

Part Three

Discerning Public and Private Spheres

A More Perfect Union

Recovering Ethics in Public Dialogue

Spoma Jovanovic
University of North Carolina, Greensboro
Roy V. Wood
University of Denver

Emmanuel Levinas is known for his compelling phenomenological account of the ethics in the proximity of self and the other against the ground of being (1998). For him, ethics is always already there. For Zigmunt Bauman (1993), this "moral party of two" is the foundation of morality in postmodernity (110). But the moral party is not automatically ethical. The temptations of the self and the limits of language can overwrite the impulse to care for the other or even respond when the other calls to us.

No sooner do we arrive at some comfort in our understanding of the self's responsibility to the other, than another other appears on the scene. This "earthquake of the third" (Bauman 1993, 112) leaves the self in a quandary. Whereas the possibility of infinite responsibility for the one other was the tall order imposed by philosophers Levinas and Bauman, now the responsibility extends to two others, then three, and then even more. The arrival of the third is the birth of politics where justice, or comparing the seemingly incomparable, emerges (Levinas 1998). If responsibility and the call of conscience are at the heart of ethics, justice is birthed with the call for fairness, equity, and care for all. Justice provides a correction to the unremitting relationship of asymmetry to the one other. Levinas welcomes justice and says, "Thanks to God that, I am another for the others!" (1998, 158).

The challenge for Levinas and his enthusiasts has been to understand the workings of the move from ethics as an interpersonal experience to justice and society as a collective encounter. Can the ethical foundation find its way into the crisis of American political communication (Zarefsky 1992)? When candidates and elected officials seek to mask the issues, mislead the public, and weaken the public's resolve, as they have, what hope remains for the possibility for change? Must self-interest, objectification, and thematization, which seem inevitable outcomes of political jockeying and the thirst for power, overwrite the ethical impulse? Perhaps not. Levinas points the way to the possibility for an ethical politics but leaves it for others to pursue the problem.

How can we bring an ethics of responsibility and care into debate, public speaking, political communication, and even presidential politics? If the ethical foundation is always present, though perhaps overwritten, how can we move "up" the platform of ethics to the center stage of politics?

The central question here is how to reclaim a Levinasian ethics in politics and justice. In this paper, we suggest that the answer may be found in the clues Levinas provides in his analysis of the love of wisdom and the wisdom of love. To make this argument, first we question the assumption that a politics based on the wisdom of love must work, as many readers of Levinas have suggested, from the bottom up, from the face-to-face ethics to justice. We explore our alternative position through a look at the phenomenology of a political speech where we see higher-order values working to move the audience to dialogue and ethics. Our case example is Barack Obama's address, "A More Perfect Union." We then discuss the use of "prophetic speaking" in Obama's remarks. Last, we invite our readers to consider whether the wisdom of love and prophetic speaking document that, in politics, justice can indeed resuscitate ethics and dialogue.

WISDOM OF LOVE AND ETHICAL COMMUNICATION

The issue before us is how to make ethics as the unconditional responsibility for the other (Levinas 1969), effective within the context of the third (the other others), of politics (Levinas 1998). Philosopher Diane Perpich (1998) has heretofore argued that justice for Levinas depends first and foremost upon the ethical relationship, for "it is ethics that endows justice with meaning" (66). Others agree that the path to justice begins with the self's ethical responsibility to the other (Derrida 1999; Bloechl 2000). Murray (2003) further notes that Levinas's philosophy of ethics, originating from the call of the other, is perhaps incompatible with a *system* (italics his) of justice, yet Murray concedes that ethics "is the necessary prerequisite and starting point for justice" (2003, 145). From this view, justice as a commitment to the multiple and often rival claims that arise from distinct others mandates that we consider our ethical obligation to the one other first to ensure that justice is indeed just. To do otherwise would be to objectify the individual. Thus, rote adherence to a rule, law or policy Levinas would argue, dismisses the self's responsibility to the other in favor of prescribed (unethical) action. This view of justice has been instructive to advance the standing of ethics in societal matters to prevent the annihilation, both literally and figuratively, of the other. But a look at the actual, phenomenological world of public political communication suggests that the ethics of the moral party of two does not transition easily to politics. Communicators see the "real world" of politics to be more Hobbesian than Levinasian. Going on the attack is part of the game. Listeners come to expect that political war suspends morality, expecting and accepting ad hominem arguments, stereotyping, and scapegoating.

Recent scholarship by Finkielkraut (1997), Burggraeve (2002) and Beals (2007) has prompted us to imagine a somewhat different possibility for recovering ethics in public dialogue. They point to Levinas's discussion of the wisdom of love as part of a new way to look at ethics and politics. From them, we see that the wisdom of love in justice—that is, following a vision set forth in, for instance, our country's founding documents—has the volitional power to call forth ethics to the situation. This directional approach of justice to ethics and dialogue may provide cause to invert our original understanding of the relationship between ethics and justice.

The wisdom-love relationship is oriented around the trope of the Greek and the Jew. The love of wisdom (Athenian) brings the rational to ethics; the wisdom of love (Jerusalem) suggests that the wise path to ethics begins with love. These two different orientations may help explain how people hold distinct cognitive frames from which their political values are situated (Lakoff 2002).

Linguist George Lakoff (2002) explains that people hone their commonsense worldviews based on family values instilled during childhood. These values in turn are enacted in our moral views and political ideologies. For example, a person who embraces what Levinas refers to as a love of wisdom might value self-reliance, individual responsibility, standards, discipline, and rewards both within a family structure and political policies. A person whose worldview is instead constructed around Levinas's reference to a wisdom of love would likely embrace values and talk that center on human rights, concern, care, help and dignity. These alternative worldviews lead us to different questions, and quite frequently, conflict.

The love of wisdom begins and ends with perfecting the self and to contain the violence that self-centeredness holds at its roots. Sound policy, social contract, and treaties are the ways to sustain peace as the absence of conflict (Lakoff 2002). But self-centeredness leads to selfism, which in turn leads to a denial of the other, racism, murder, and hate. It births the idea of unlimited personal freedom. That breeds conflict among individuals and wars among countries. Segregation, conflict resolution, and empty legislative action bring moments of peace as interruption to hostility and war.

The wisdom of love starts from the love of one's neighbor. It is other-centered and founded on responsibility *for* the other as a condition of being *with* the other (Levinas 1969). It interrupts the self and evokes "Thou shall not kill." The other is conceived as the moral center, to whom we owe everything (Lipari 2003). Instead of controlling original evil, the wisdom of love seeks to return to original good. There is freedom, to be sure, but it is difficult freedom carved out from the ground of social responsibility (Lakoff 2002).

Operating from the wisdom of love suggests how we might bring ethics and dialogic responsibility into the world of justice and politics. Who would have thought that a presidential candidate would show us the way? Who would have guessed that he would do that by turning the problem back on its head? Yes, justice is founded on ethics, but just values can turn back upon the system to re-birth ethics that have been lost.

A SPEECH OF HOPE AND UNITY

On March 18, 2008, Senator Barack Obama spoke to the nation on the subject of race and in doing so directed our attention to the need for a new politics in America that could embrace a vision of hope and unity (Lakoff 2008). In Philadelphia's Constitution Hall, Obama stood before the American people to address embarrassing sound bites culled from sermons by his former pastor, the Reverend Jeremiah A. Wright, Jr. It was a speech resounding with compassion and reflective of a deep concern over the cultural divides in the nation. Obama embraced the opportunity to speak of the contradictions that persist in American life at the same time that he asserted a faith that ours can indeed be "a more perfect union" than we have to date realized. Called "masterful," (Pitts 2008), "unprecedented" (Bloice 2008), and "relevant" (Lerner 2008), Obama's speech presented the challenge before us in the struggle for social justice; he urged people to assert their convictions and importantly, their actions, to work for a democracy that speaks to the enduring values of love of our neighbors, wisdom of the lessons of history, and insistence for justice in every realm of our lives.

Obama's speech surprised us. Rather than doing what we have come to expect from our politicians—explaining away a problem, disavowing a sentence, discrediting a person, and then moving the conversation quickly back to standard political platform sound bites—Obama introduced us to a different political discourse. At a time when "political communication is *assumed* to be inauthentic, to mask the real issues, and to pervert rather than strengthen the exercise of judgment" (Zarefsky 1992, 414), Obama spoke instead of the strength and promise of our democracy "... based on empathy, on the bonds of care and responsibility that link us together and make us a nation" (Lakoff 2008).

Obama set a context for his 37-minute speech that originally aired on CNN by opening with powerful words culled from our country's founding documents:

> "We the people, in order to form a more perfect union." Two hundred and twenty-one years ago, in a hall that still stands across the street, a group of men gathered and, with these simple words, launched America's improbable experiment in democracy . . . the document they produced was eventually signed but ultimately unfinished. It was stained by this nation's original sin of slavery, a question that divided the colonies . . . (2008)

Newspaper columnists, television analysts, pundits, and citizens across the country were buzzing of the news of Obama's speech in the days that followed. Collectively, they noted the speech for its "teachable moment" to discuss matters of race and other substantive concerns in terms not just black and white (Pitts 2008). What Obama accomplished in this rhetorical feat was to hold up the nuanced ways in which the conversations about race most often seek to divide us when in fact complex problems in education, health, and labor affect all of us directly, regardless of race, in the quality of our interactions and in the results of our decisions at the personal and political levels. He leaned on the values as put forth in our country's founding documents to ground our discourse in a new paradigm of hope—hope that we could see the need for all of us to do our part, as had our forefathers. Obama (2008) reminded us how American principles have been put into action by the people to protest on the streets, initiate legal action to challenge injustices, participate in acts of civil disobedience, and "continue the long march of those who came before us, a march for a more just, more equal, more free, more caring and more prosperous America."

Obama's task included an affirmation of his relationship with his friend and former pastor. Obama refrained from using words to demonize, as had been the route chosen by many others, instead choosing to uplift Reverend Wright's contributions to humanity, for which he had been long recognized.[1] The speech's power, however, ultimately came from his longer commentary on the entrenched issues of racism and disenfranchisement in this country. Obama recognized that any meaningful change in our political structure relies on the efforts of all of us to initiate dialogue and then insist that our political leaders respond to the messages of care that citizens want. In this way, Obama illustrated what communication ethicist Ken Chase defined as the ethical rhetor more than a decade earlier:

> In an ethically reconceptualized rhetoric [following Levinas], the rhetor is one who responds to the competing obligations. The rhetor therefore, will revitalize a personal sensitivity to justice and cultivate insights into the Other . . . The rhetor, then, is a medium, a crossing guard, the point of convergence and simultaneously, of divergence for the trafficking of human calls (1994, 227).

Obama used his speech to communicate his own responsibility and evoke responsibility as well from the audience.

In stark contrast to the politics we have come to know as disingenuous, Obama's speech took us back to American political values and political morals that reflect a deep concern for the values of cooperation and collective effort, values that are the foundation of our very freedom (Barber 2007). Obama beseeched us to reclaim our democracy from corporate greed, media mania, and elite politicians by relying on our compassion and respect for the other and by speaking up about uncomfortable truths rather than retreating from them (West 2004a). Obama's speech, in word and tone, reflected the dialogic I-Thou relationship Martin Buber so persuasively insisted was a requirement for civil society (1955). As Buber would have it, Obama made himself fully present to the audience, focused on discussing difficult issues by plunging into the heart of what he had seen ailing Americans. In dialogic fashion, Obama deconstructed common stereotypes to reveal what drives the words and actions of ordinary people and media-made celebrities.

From Justice to Ethics

Obama did *not* apply ethics to politics but instead showed through his communication how politics and a concern for justice could lead us to ethics. Obama highlighted how justice is embedded in the constitution in such critical, foundational values as equality, liberty, and union. With these values to guide us, our interactions unavoidably reflect a care for others, not a demonization of them. With these values firmly established and discussed, our communication moves away from blame and retribution for all that remains unjust to ideals, hopes, and thoughts of a transformative moment when the needs of others becomes the focus of our genuine concern.

Pluralism demands that difference be recognized and regarded as the very basis of a democracy's social and political agenda (Mouffe 1999). Of immeasurable importance in a just world is the matter of how we acknowledge others. Obama used communication to bind us as Americans, at the same time that his speech pointed out the distinguishing features of our everyday lives. This move attended to the contingencies of pluralism not as a force to be subdued, but as a promise for peace. For Levinas, justice is achieved when it is humane through and through by recognizing and granting direct discourse as a response to the universals in which justice is always situated (1969). That is, justice arises through connecting, not distancing or neutralizing strategies. We see this in Obama's speech that pointed us to the larger landscape of justice by illustrating with specific issues, names, and stories the political challenges and ethical obligations we face.

Phenomenologically Speaking

How, phenomenologically, did Obama use rhetoric to direct our attention to foundational political, religious, and moral values so as to raise our consciousness to the fundamental issues of race in America? We take the phenomenological method to be one that examines action in the world as it occurs in the moment, mining for the evocative moments and the concurrent sense making (Jovanovic and Wood 2004). We view this address as a speech act, spoken from the heart, in which the speaker/writer turned away from "campaign rhetoric" to deal with a genuine personal crisis and to reflect on a major issue in the lived experience of Americans. It featured ". . . the concept of ethical responsibility through intersecting levels of phenomenological description . . . [and] hermeneutic phenomenology of the prophetic vocation of the subject (Bergo 1999, 210).

Obama anchored his speech in an analysis of the fundamental American values articulated by the founders in two documents, signed 11 years apart. The Declaration of Independence took us to self-evident truths—all men are created equal with the unalienable rights of life, liberty, and the pursuit of happiness. The founders made that real through the development of the Constitution with the significance attributed to the ideal of a more perfect union formed for justice, defense, and domestic tranquility to promote the general welfare and to secure liberty for themselves and their posterity.

Woven among the exposition of slavery as the American original sin, the dimensions of his relationship with Reverend Wright, hidden black anger, hidden white anger, distractions that keep us from addressing race, the need for change, and the conviction that it can happen, were 12 paragraphs in the speech out of 60 that were dedicated to clear, articulated values—political and religious.

The overriding values in Obama's remarks were unity over divisiveness, perfecting the union, *e pluribus Unum,* equal citizenship, liberty, and justice. He called for what all great religions demand—to do unto others as you would have them do unto you; to be your "sister's keeper." Then, between the lines, were articulations of care and responsibility for others who need health care, education, justice, and jobs. Obama's speech ended with the story of two people, one of whom would concur with Levinas that ". . . the other is always first, not last" (Lipari 2003, 9).

From the vantage point of the love of wisdom, this was not a prudent speech. A shrewder candidate might have ended the speech at paragraph 23, about a third of the way through. Obama acknowledged this when he next said "I suppose the politically safe thing to do would be to move on from this episode and just hope that it fades into the woodwork" (2008). No, he moved on instead to say, "But race is an issue that I believe this nation cannot afford to ignore right now." A sensible speaker would have minimized the violence that could be done to him, but Obama turned instead to the values that flow from the Levinasian wisdom of love—unity, equality, justice, care, responsibility, and difficult freedom.

Obama called for the presidency to build upon such values and lamented the ethical deterioration that had led to politics of attack, division, and sound bites. Here students of Levinas would likely expect the speaker to construct his speech on the foundation of love found at the level of the face-to-face interaction. He could call for the application of ethics to political speeches and acts, exhorting the candidates to behave responsibly. Instead, Obama built to the face-to-face with a different kind of speaking that led from the values that must evolve in the world of "the third," of politics and justice. Those values can serve to remind us, when we are egocentrically overcome by thematizing and objectifying the other, of who we were when we started to perfect our union.

To say, though, that Obama got to ethics through the high values of justice at the founding of America does not fully address the question of whether he worked from love or wisdom. He could, in fact, have used a traditional logical approach by placing the values as warrants for his conclusions. But there is more to his speech than argument. We need now to address his very style of prophetic speaking.

Prophetic Speaking

Prophetic speaking, with its roots in Hebrew scripture, proclaims that to be human is to be kind to the stranger. That is, humanity begins as a response to the cries of another (Hyde 2001; Levinas 1998; West 2004a). Political philosophers and political candidates who embrace this tradition in their writings tell us that democracy is always going to be a project that asks, what is the relationship between the public and its most vulnerable citizens (Levinas 1998; West 2004b)?

Prophetic speaking *is* speaking from the wisdom of love, and for Levinas that means to put the self into the fray to confront injustice that afflicts not the self but instead the other (Beals 2007). It bears repeating that in his speech, Obama intentionally put himself into the heat of the debate surrounding race:

> But race is an issue that I believe this nation cannot afford to ignore right now. We would be making the same mistake that Reverend Wright made in his offending sermons about America—to simplify and stereotype and amplify the negative to the point that it distorts reality.

We cannot, in other words, retreat from the difficult subjects of our time or pretend they are not as significant as reality reveals. A February 2008 report to the United Nations Committee for the Elimination of Racial Discrimination confirms that structural racism is not a figment of the Black man's imagination. The report says, ". . . prosecutors seek the death penalty in 70 percent of cases involving Black defendants and White victims, yet only in 19 percent of the cases involving White defendants and Black victims" (Menendian, Spencer, Knuth, Powell, Jackson, Fajana, Grant-Thomas, Reece, Paterson and Rapp 2008, 5). To be Black in America is to suffer unfair treatment in the courts, inferior education in schools, and lack of access to economic opportunities for wealth. The response, Obama said, has created the conditions in which shame grew, frustration mounted, and violence erupted. The conditions were of our own making in policies of the government, and so Obama reminded us that ". . . the anger is real; it is powerful; and to simply wish it away, to condemn it without understanding its roots, only serves to widen the chasm of misunderstanding that exists between the races" (2008). He added, "In fact, a similar anger exists within segments of the white community" who do not feel they have been privileged because of the color of their skin. For Obama, the idea and actuality of connection undergirded his speech, his identity, and his political outlook:

> I am the son of a black man from Kenya and a white woman from Kansas. I was raised with the help of a . . . white grandmother. I've gone to some of the best schools in America and lived in one of the world's poorest nations. I am married to a black American who carries within her the blood of slaves and slave owners . . . I have brothers, sisters, nieces, nephews, uncles and cousins, of every race and every hue, scattered across three continents . . . It's a story that hasn't made me the most conventional candidate. But it is a story that has seared into my genetic makeup the idea that this nation is more than the sum of its parts—that out of many, we are truly one (2008).

Obama embodies the radical alterity about which Levinas insists is the basis of sociality and the promise of a future that regards difference with awe (1969). From Levinas, we learn that community is constructed through prophetic language (Bergo 1999).

Prophetic speaking is forward looking in its attempts to unshackle those who have been left behind; it is also historically situated, recognizing that some people have been subsumed by the unjust or unethical conditions of the past. Taken together, prophetic speaking recognizes what some have tried to keep invisible (the pain, the suffering) in hopes of realizing an existence where being for the other becomes the default social experience. What this way of speaking requires, as Obama illustrates, is to be concerned not only with those to whom we are close but also to all the others who comprise our larger community:

> . . . justice remains justice only, in a society where there is no distinction between those close and those far off, but in which there also remains the impossibility of passing by the closest. The equality of all is borne by my inequality, the surplus of my duties over my rights. The forgetting of self moves justice (Levinas 1998, 159).

Levinas's view is that "the ethical shows itself as the prophetical: this means that ethics expresses a vocation for humanity" (Ciaramelli 1991, 92). For Obama, the prophetic voice emerged again and again, most poignantly perhaps when he spoke of the persistence of poverty and crime that accompany the lived disappointments of people unable to capture a piece of the American pie:

> But for all those who scratched and clawed their way to get a piece of the American Dream, there were many who didn't make it—those who were ultimately defeated, in one way or another, by discrimination. That legacy of defeat was passed on to future generations—those young men and increasingly young women who we see standing on street corners or languishing in our prisons, without hope or prospects for the future (2008).

Obama granted blacks and whites, and all Americans, their justified feelings of abandonment by a country where jobs are shipped overseas, oil and food prices escalate, and more dual wage earning families are having by necessity to choose between food or medicine rather than having both. It is this recognition, that we all are implicated in the struggle for a common welfare that reveals Obama's interpretation of the wisdom of love as a rallying cry for Americans:

> . . . I have asserted a firm conviction—a conviction rooted in my faith in God and my faith in the American people—that working together we can move beyond some of our old racial wounds, and that in fact we have no choice if we are to continue on the path of a more perfect union (2008).

Obama's propheticism calls the community to *be for* the other (Bergo 1999).

Obama's speech training no doubt came from his legal background, his community service, and his religious roots in the Black Christian church. In that church, the prophetic tradition is used to reflect a worldview that is a response to the perpetual economic domination and profound racism that originally confined Blacks to their own churches for spiritual relief (West 1988). Black Americans found hope in sermons espousing moral revelation and redemption to dislodge the inevitable despair that draped their lives. Obama leaned on this tradition to ask Americans of every race, "Let us find that common stake we all have in one another, and let our politics reflect that spirit as well" (2008).

In fact, Obama warned us that if we do not opt for the prophetic, we are destined to retain instead a self-focused culture that fears the other, pounces on the other's mistakes, and blames the other for our own misfortunes. To choose that path, said Obama, guarantees that we remain in a "racial stalemate" that becomes a metaphor for a nation polarized.

CONCLUSION

When Obama prepared to speak on that day in March 2008, his national audience expected him to disavow his association with Reverend Wright, who had made comments labeled un-American. That would have been the expedient thing to do and what Americans have come to expect as prudent political strategy. It would have been the reasonable path to follow. Instead, Obama delivered a breathtaking speech that acknowledged the complexities of outrage steeped in historical wrongs, wrongs that remain among the major issues of our time. To do that, Obama turned to the promise and hope of our United States Constitution to note that the unfinished business our forefathers left for us—the inequities surrounding race and economics—remain in the twenty-first century. The time, he said, had come to face those issues squarely by feeling the pain and the struggle of those disenfranchised by the political systems in place and asserting a new, collective vision to right the lingering wrongs. Obama aimed toward a transformative moment.

> Speaking of a transcendence beyond the essence is the losing of your own breath, a sort of expiration, or at least, a respiration that is not owned but received and in which a wholly different spirit than that of ontology can—enigmatically be heard (Peperzak 1997, 90).

Obama's speech was a response to the other—to Reverend Wright and to the American people—that sought to infuse the wisdom of love into the arena of reason.

In this presidential campaign, Obama proclaimed that race would be discussed rather than dismissed. Inadequate access to medical insurance, education, and jobs would also be the subjects of important dialogue he would initiate. Obama stepped into the middle of problems that we had not solved heretofore, to offer not a solution but a commitment to work *with* the citizens of the country to discuss and decide how to chart a course reflective of the core tenets of freedom, care, and responsibility around which the dream for this country was formed.

To be sure, this speech by presidential hopeful Barack Obama served two fundamental purposes. First, he faced a rhetorical crisis engendered by sound bites taken from sermons by Obama's former minister, the Reverend Jeremiah Wright. The candidate needed to disown the minister's remarks while reaffirming the positive qualities of their long association and the minister's service to his family, congregation, and nation. Second, while he was expected to stop there, Obama seized the opportunity to re-establish race in America as central to the country's next steps in fulfilling its dream. This paper addresses a third, even more fundamental construction behind the first appearances of the speech. Barack Obama asserted the challenge that the time has come to recall the fundamental values embedded at the heart of the American project. Those values are statements about the essential dream behind politics and justice in America. In the best Levinasian sense, they grew through the wisdom of love from ethical values at the level of the face to face. Unity, equality, and care for the general welfare, as well as opportunity and freedom, are extensions of the acknowledgment, proximity, responsibility, care, and difficult freedom about which Levinas speaks. But the deeper crisis addressed by the speech is that those values have been lost.

American politics has become selfist. "*I* am the best" and "*I*" will solve your problems for you" are campaign mottos adopted by virtually every candidate. American politics has become divisive. "*They* are immoral" and "Where is *your* American flag lapel pin?" are intended to suggest that the other candidate or political party is not good enough or really, even good. Argument over issues derived from values has been replaced by "gotchas." Having to address Reverend Wright's beliefs instead of his own served as just another example of the distractions that Obama set out at the beginning to avoid.

So in the background yet foreground of the speech was an attempt to recollect the political and religious values that could lead us back to ethical politics and ethical relationships that are at the heart of the dream. In our view, Obama succeeded in his attempt to do that and, in that, reestablished our political values, recalled our spiritually based role as our brother/sister's keeper, established an agenda for conversations about race, and led us then back to ethics and dialogue.

This last move was not as obvious as others, but phenomenologically and theoretically it was perhaps the most interesting. Can ethics lost, subverted by egocentrism and the love of wisdom that serves it, be regained? Yes, but only, we would have thought, by trying to return to the bedrock of the face to face hoping that that would "trickle up" to the world of politics. In the midst of our realization that the hopes for that are bleak, this candidate has set the problem on its head starting at the top and moving to ethics and dialogue.

Obama ended the speech with a story. It is a story that shows we can unite, and we can be ethical, and we can engage in Levinasian dialogue. In the story he recalled a white woman named Ashley who organized his campaign in a town in South Carolina. The volunteers went around the room to tell how they got involved, and Ashley's story was a heart-rendering one that underlined that poverty affects us all.

> Anyway, Ashley finishes her story and then goes around the room and asks everyone else why they're supporting the campaign. They all have different stories and reasons. Many bring up a specific issue. And finally they come to this elderly black man who's been sitting there quietly the entire time. And Ashley asks him why he's there. And he does not bring up a specific issue. He does not say health care or the economy. He does not say education or the war. He does not say that he was there because of Barack Obama. He simply says to everyone in the room, "I am here because of Ashley." "I'm here because of Ashley." By itself, that single moment of recognition between that young white girl and that old black man is not enough. It is not enough to give health care to the sick, or jobs to the jobless, or education to our children. But it is where we start. It is where our union grows stronger. And as so many generations have come to realize over the course of the two hundred and twenty-one years since a band of patriots signed that document in Philadelphia, that is where the perfection begins. (2008)

Poulos (2008) would surely see this story that Obama told as the beginning of an "accidental dialogue" where the opportunity for transformation was recorded phenomenologically, and born from imagination, courage, and narrative conscience.

Within the crisis of American political communication, Obama's speech offers a ray of hope that the wisdom of love can guide us toward the responsibility and care we so desperately need to reinvigorate civic involvement. The answers to the complex problems of our times are not to be found in glib formulas, gas-tax vacations, or the policies of leaders alone. Instead, a rhetorical focus on justice for all in speeches and in dialogue, provides a route to resuscitating an ethical public climate in which all are honored. Justice will be just, Levinas (1998) would surely say, when people, black, brown, yellow, red, and white, young and old, rich and poor, rural and urban, in talk and in deed, enter into relationships that extend beyond even the scope of their expedient intentions to care for and *be* for one another.

REFERENCES

Bauman, Zygmunt. 1993. *Postmodern Ethics.* Oxford, UK: Blackwell.

Beals, Corey. 2007. *Levinas and the Wisdom of Love.* Waco, TX: Baylor University Press.

Bergo, Bettina. 1999. *Levinas Between Ethics and Politics: For the Beauty that Adorns the Earth.* Dordecht, Netherlands: Kluwer Academic Publishers.

Bloechl, Jeffrey. 2000. *Liturgy of the Neighbor: Emmanuel Levinas and the FReligion of Responsibility.* Duquesne, PA: Duquesne University.

Bloice, Carl. 2008. "Obama race speech analysis by 14 BlackCommentator.com editorial board members." *The Black Commentator* 269 (March 20). www .blackcommentator.com (accessed March 25, 2008).

Buber, Martin. 1955. *Between Man and Man.* Trans. R. G. Smith. Boston: Beacon Press. (Original Work published 1947.)

Burggraeve, Roger. 2007. *The Wisdom of Love in the Service of Love.* Milwaukee, WI: Marquette University Press.

Chase, Kenneth. R. 1994. "Rethinking rhetoric in the face of the other." J. A. Jaksa's *Conference Proceedings of the 3rd National Communication Ethics Conference.* Annandale, VA: Speech Communication Association, 220–229.

Ciaramelli, Fabio. 1991. "Levinas's Ethical Discourse between Lndividuation and Universality." In *Re-reading Levinas*, ed. Robert Bernasconi and Simon Critchley, 83–105. Bloomington, IN: Indiana University.

Derrida, Jacques. 1999. *Adieu to Emmanuel Levinas.* Trans. P-A. Brault and M. Naas. Stanford, CA: Stanford University.

Finkielkraut, Alain. 1997. *The Wisdom of Love.* Lincoln, NE: The University of Nebraska Press.

Hyde, Michael J. 2001. *The Call of conscience: Heidegger and Levinas, Rhetoric and the Euthanasia Debate.* Columbia: University of South Carolina Press.

Jovanovic, Spoma and Wood, Roy V. "Speaking From the Bedrock of Ethics." *Philosophy and Rhetoric* 37(4): 317–334.

Lakoff, George. (2008, 2004): "What Made Obama's Speech Great." (March 26) www.alternet.org/election08/80549/ (accessed May 1, 2008).

——. 2002. *Moral Politics: How Liberals and Conservatives Think.* 2nd ed. Chicago: University of Chicago Press.

Lerner, M. (2008). "The Obama Phenomenon." *Tikkun.* (March). www.tikkun.org/magazine/tik0803/frontpage/phenom (accessed April 24, 2008).

Levinas, Emmauel. 1969. *Totality and Infinity.* Trans. Alphonso Lingis. Pittsburgh, PA: Duquesne University. (Original work published 1961.)

——. 1998. *Otherwise Than Being: Or Beyond Essence*. Trans. Alphonso Lingis. Pittsburgh, PA: Duquesne University. (Original work published in 1974).

Lipari, Lisbeth. (2003). "*Recognition and the Other*." Paper presented at the International Communication Association Annual Meeting, San Diego, CA.

Mouffe, Chantal. 2000. "Which Ethics for Democracy?" In *The turn to ethics*, ed. Marjorie Garber, Beatrice Bannssen and Rebecca L. Walkowitz, 85–94. New York: Routledge.

Murray, Jeffrey W. 2003. *Face to Face in Dialogue: Emmanuel Levinas and (the) Communication (of) Ethics*. Lanham, MD: University Press of America.

Obama, Barack. 2008. "Remarks of Senator Barack Obama: 'A More Perfect Union.'" *CNN-TV*. www.youtube.com (accessed March 19).

Peperzak, Adriaan Theodore. 1997. *Beyond: The Philosophy of Emmanuel Levinas*. Evanston, IL: Northwestern University.

Perpich, Diane. "A singular justice: Ethics and politics between Levinas and Derrida." *Philosophy Today*, SPEP Supplement, (1998): 59–70.

Pitts, Steven. 2008. "Obama race speech analysis by 14 BlackCommentator.com editorial board members." *The Black Commentator* 269 (March 25). www.blackcommentator.com (accessed March 25, 2008).

Pitts, Jr. Leonard. 2008. "Obama provides a lesson on how to talk about race." *Greensboro News & Record* (March 24), A9.

Poulos, Christopher N. "Accidental Dialogue." *Communication Theory* 18 (2008): 117–138.

West, Cornel. 2004. *Democracy Matters: Winning the Fight against Imperialism*. London: Penguin.

——. (2004). "*Socratic and Prophetic Approaches to Democracy (speech)*." Hollywood, CA: Barnsdale Art Park. (December 4) www.knitespace.com (accessed May 2, 2008)

Zeleny, J. and Nagourney, A. (2008). "An Angry Obama Renounces Ties to his Ex-Pastor." *New York Times*. (April 30). www.nytimes.com/2008/04/30/us/politics (accessed June 10, 2008)

Zarefsky, David. "Spectator politics and the revival of public argument." *Communication Monographs* 59 (1992): 411–414.

NOTES

1. Five weeks later, after a series of more incendiary televised comments by Reverend Wright just before the Democratic primaries in Indiana and North Carolina, Barack Obama renounced his ties to his ex-pastor. Obama noted that Rev. Wright's divisive comments contradicted "everything that I'm about and who I am" (Zaleny and Nagourney, 2008, ¶ 11).

The Crisis Fallacy

Egoism, Epistemology, and Ethics in Crisis Communication and Preparation

Pat J. Gehrke
University of South Carolina

Crisis studies are in a crisis. More appropriately, we might say that contemporary crisis studies have come to a surprising point in which their increasing precision and preparedness for crises has actually diminished their utility and, in many cases, reduced crisis communication to little more than corporate image repair. Based on a review of crisis communication and crisis preparation literature, this essay argues that our current definition of crisis reflects an egoistic orientation and undermines our ability to both prepare for and respond to crises. The re-definition of crisis offered in this essay was derived by analyzing a collection of major texts on crisis communication[1] with special attention to their treatment of case studies and examples. In doing so, this definition depicts how crisis is operationally defined in actual practice, rather than in abstracted statements of principle. The definition of crisis derived in this study is not flattering to contemporary crisis scholars and practitioners, but perhaps even more worrisome is the combination of an egoistic orientation with a set of epistemological blinders that undermine preparation for unforeseen high-impact events by relying upon preparation of response plans and adopting a fallacious relationship to the future. These knowledge problems exacerbate the problems of egoism found in crisis studies while, at the same time, making it impossible for the dominant approaches to crises to serve even their narrow egoistic purposes.

As salve to this malady, organizations ought to adopt a disaster orientation with its broader hermeneutic rather than the narrowly egoistic crisis perspective. In so doing, they must also move from a scenario-response planning model into a more sustained organizational culture that builds coalitions and engages a broad range of stakeholders. In those coalitions, rather than planning for the predictable, the emphasis ought to be placed on using highly improbable and imaginative scenarios to cultivate an adept and adaptable capacity for judgment. Such a shift in orientation, culture, and judgment would not only help to prepare organizations and individuals better for the impacts of the highly improbable but also, as a prerequisite to being so prepared, diminish the egoistic orientation that exacerbates crises and encourages pathological behavior in organizations.

WHAT CONSTITUTES A CRISIS?

Robert Ullmer, Timothy Sellnow, and Matthew Seeger's definition is a good representative of the field, describing crisis as "coming as a surprise, posing a serious threat, and forcing a short response time" (2006, 6). Elsewhere, the same authors have defined crisis as "an unusual event of overwhelming negative significance that carries a high level of risk, harm, and opportunity for further loss" (Seeger, Sellnow, & Ulmer 2003, 4.). Their definitions share the importance of surprise, threat, risk, and response, which we find in most formal definitions of crisis in the literature. However, these definitions do not reflect certain phenomenological components that better describe how crisis communication literature constructs and engages events, in addition to how crisis consultants and communicators frame their practices.

Reviewing our contemporary discussions of crises and crisis theories generated the following more complete definition of crisis: *the egoistic experience of a constructed event that comes as a surprise and poses a moment of judgment perceived to be of heightened importance.* This definition, derived from the practices and discourses of crisis communication scholars and practitioners, contains five components:

1. constructing an event,
2. experiencing that event egoistically,
3. being surprised by the event,
4. calling for judgment and,
5. perceiving the event as holding heightened importance.

This definition differs in a few significant ways from those commonly found in major crisis communication texts. Most important among these elements is the second: crises are differentiated from other kinds of events, such as disasters, by an egoistic orientation. As a result of this orientation, a crisis approach to events often encourages egoistic behaviors in organizations and persons who engage in preparedness for, response to, and communication about critical moments of judgment. Such egoism pervades the crisis orientation to such a degree that we can see its influence in construction of, surprise at, judgment about, and evaluation of such events. Additionally, the proposed re-definition of crisis shifts the analysis from the quality and content of the event itself and toward how we construct and encounter the event, which offers an improved understanding of how responses are both prepared and justified. To clarify this definition further, the next section explicates each component, demonstrating its operation in commonly used case studies in crisis communication.

Crisis as Constructed Event

Crises are constructed as single (and singular) events. Isolating a crisis as an event does not mean that it occurs in a single instant, for an event can be manifest over a sequence of time. For example, product tampering and product defects usually manifest over days, weeks, or even months, but the whole of that series of time is understood and experienced in crisis communication as an event. The 1982 Tylenol murders involved seven victims who died over a period of days. Likewise, when one DC-10 aircraft crashed during takeoff due to a simultaneous failure of engine power and slat control it was an accident. When four did so in four years, McDonnell-Douglas experienced a crisis. Even crises that appear to occur in a moment, such as the collapse of the Crandall Canyon Mine in 2007 or the explosion of the space shuttle Challenger in 1986 are crises by virtue of a sequence of actions, decisions, and instances over time that come together to form the event as a crisis. In the case of the Crandall Canyon Mine there were previous safety concerns, controversies over the strategy of "retreat mining," and a history of union disputes, which all composed the crisis event, regardless of whether or not they might have been a part of the disaster that was the mine collapse.

To say that crises are constructed is not to say that things like plane crashes, mine collapses, and product tampering are built solely through rhetoric. Instead, we ought to recognize that those events take on crisis status when they are spoken about and understood in a particular way. What makes one plane crash a crisis and another an accident or transforms a mine collapse from tragedy to crisis is its contextualization, its framing, or the way the event comes to be depicted and interpreted. Like all rhetorical construction, when a crisis event is constructed there is never merely one agent nor a single dimension of agents involved in the construction. While Robert Murray of Murray Mining might wish to believe that the crisis for his company was fabricated out of the Crandall Canyon Mine collapse by the union and media outlets, the truth of the matter is that his own rhetoric and that of many others prior to the collapse and after made that collapse a crisis rather than an accident or a tragedy. For an event to be rhetorically constructed as a crisis the qualities of the event must be interpreted as critical by not only the organization or persons who might be called to respond to the crisis but also by others who compose the broader ecology in which that organization exists. Certainly that might include a union or media outlets, but in many cases it also includes customers, regulators, law enforcement, shareholders, employees, community groups, and others as well.

The event transforms from disaster to crisis when enough people begin to understand the result in crisis terms, in terms that relate the event to responsible persons or organizations. The DC-10 plane crashes became a crisis for McDonnell-Douglas when interpretations began to take hold that understood the crashes in relation to possible flaws with the DC-10 itself, which may have been manifest by the simple rhetorical act of linking four crashes together by virtue of the same make and model of plane. The Tylenol murders were a crisis for Johnson & Johnson as soon as there was a broad conversation about the connection of these deaths to taking the pain reliever. In these cases, the rhetorical construction is simple, following along the nomenclature of a product, playing off the power of brand recognition. Even in such simple cases, however, the formative construction is clearly a matter of connecting perceptions via language to build a broader potential result. Implicit statements about the organizational *ethos* and issues of potential liability are bound up with conversations about why four of this specific model of plane all suffered similar catastrophic failures or why only bottles of this brand of pain reliever caused seven deaths. At the moment such questions are raised a crisis begins to emerge on the scene, a crisis that previously may have been only possible or inchoate, even if all the instances and occurrences were in the past. In short, what makes a crisis a crisis is that it has been rhetorically constructed as a crisis for some person, group, or organization, who may encounter some future negative outcome if that rhetorical construction takes hold in the opinions and beliefs of people who can have impact upon that person, group, or organization.

Crisis as Egoistic

Observing how crisis studies and crisis communication literature discuss crises not only demonstrates that crises are rhetorically constructed events but also displays a particular type of rhetorical construction that frames how individuals and organizations respond to crises. We know that a crisis is not objectively a crisis, but instead it is a crisis *for* someone, a crisis *for* a person, an organization, a community, *et cetera*. Even the broader term "disaster" can be understood as having this kind of subjective bias. Is a fire a disaster? That entirely depends upon *for whom* you are asking the question. For a homeowner or a human community, it most often is, though not always. For nature, the global environment, the long-term health of an ecosystem, it depends a great deal on the fire. Some species and ecologies depend upon occasional fires. However, unlike disasters, discourses of crisis give this subjectivism a much narrower scope and display instead what is best described as egoism.

Egoism should be distinguished from egotism, for the two are often confused. Egotism is thinking too much of oneself, having an exaggerated sense of one's own value or importance, and is what we commonly mean when we refer to someone having too much of an ego, being egotistical. I do not think one ought to levy this charge against crisis discourse. However, egoism does not require such an inflation of one's own position but instead implies only that one is, in short, selfish. Egoism is looking out solely or overwhelmingly for one's own self-interest, valuing one's own well-being first and foremost. Egoism is usually contrasted to altruism, and many egoists argue that true altruism does not exist, as even altruistic acts are conducted because they are in one's own self-interest (producing pleasure, fostering reciprocation, etc.). More fundamentally, egoism not only implies actions chosen on the basis of self-interest but also deploys an epistemology and hermeneutic that understands the meaning of events, people, animals, and objects as what they mean for (how they have value for) the self. Ayn Rand's objectivism is probably the most influential specific brand of egoism, most explicit in her *Virtue of Selfishness*, and has been used to defend a number of self-serving communication practices including commercial advertising (e.g., Kirkpatrick).

When we look at discourses of crisis, especially in the crisis communication literature, we see this kind of egoism at work. The one who declares an event a crisis both rhetorically constructs the existence of that event as crisis and narrows the meaning of that event down to its (usually negative) value for the person or group said to be facing the crisis. Is your loss of market share a crisis? Not for your competitors, and probably not for your customers. If one looks, for example, at the crisis literature on the Tylenol tampering cases, the crisis is constituted as a crisis *for* Johnson & Johnson. When scholars define the events—the tampering, the deaths—as a crisis, they most commonly mean that there is an organization, in this case Johnson & Johnson, for whom those events posed a crisis. The crisis is not described as a crisis for the victims, the dead, their families, or a public, but instead is a crisis for the company or the organization. Indeed, all of the other affected parties in such events are reduced in crisis literature to their relationship with the organization or to events that have already been defined in relation to the organization. Publics become customers or possibly antagonistic citizens who would lobby for more regulation. Victims become litigants, lightning rods for negative publicity, media images that would hurt the organization, or otherwise are reduced to their role in the organization's egoistic hermeneutic.

This differs substantially from the literature on disaster and disaster communication. Disaster studies and disaster communication literature surely engages in the rhetorical definition and assignment of events as disasters, and its biases are frequently anthropocentric and sometimes nationalistic or even provincial. A disaster is still never a disaster until someone can succeed rhetorically in declaring it such (which is not the same as a formal declaration), and the disaster is also still a disaster for someone (or in its broadest and most liberal uses, for something). An earthquake is merely seismic activity until someone can define it by its relation to some population, some interest, or some value. However, the value orientation of disaster literature tends to be much broader, with disasters focusing upon their damage to a community, a state, a nation, humanity, or even the planet (c.f. National Research Council, Committee on Disaster). In this regard, disaster studies and disaster communication literature tend to display a broader concern for social good as the objective (with "social good" taking varying degrees of provincialism or cosmopolitanism), while crisis literature and especially crisis communication tends to take a much narrower concern for a single organization or agency (which is most commonly a corporation), making social good meaningful only to the extent that it is beneficial to that organization.

Of course, we do see appeals to a broader good in crisis literature when organizations are admonished to be good citizens, but when we look to the crisis communication literature we almost always find that the importance and value of the broader social good is itself defined by egoism. The arguments made in favor of good corporate citizenship are that the organization's own self-interests are best served by being socially responsible. Not to say that such programs cannot do a great deal of good—many do. However, the good such programs do is neither a refutation of the egoism of the organization nor even an exception to that egoism. Instead, it is usually an investment that will be measured not by the good it does for a broader community but by the benefits it brings to the organization in the form of publicity, public relations, or fundraising.

Crisis as Surprise

Regardless of whether we later come to believe that a crisis event was entirely predictable or even that the organization had warning about its possibility, at the time of the crisis it is experienced as a surprise. Most commonly, as Seeger, Sellnow, and Ulmer write, a crisis begins with "some dramatic and surprising trigger event" (2003, 4), but sometimes a crisis can also be the surprise of a series or sequence of events. Certainly the shuttle Challenger's explosion was a surprise, but in the case of the DC-10 crashes it may be hard to distinguish when surprise took hold. Was it the second crash, or the third, or the fourth? It may be hard to determine exactly when the surprise arrived, but a crisis becomes crisis because it was unexpected or at least is constructed as unexpected. Later, or even as part of the establishment of the crisis itself, we may come to learn that a person or even the organization did know that something was going to happen, or could happen, or perhaps should have known, but the event itself will retain the qualities of surprise, of an unexpected occurrence, even if the surprise shifts to surprise that the event occurred even when it was already predicted. Thus, surprise may arise not only when events were unexpected but also from events that were expected, when prediction or foresight ought to have forestalled the events from actually occurring. In short, surprise may be brought by the failure of a party to take action to prevent what was predicted or expected.

In addition to the arrival of the unexpected, surprise also contains the element of uniqueness. Whatever moments or instances have been constructed together to form the body of the crisis, that constructed event must have a certain degree of uniqueness to formulate as a surprise. If a disaster is *de rigeur* or has become rote, such as the annual flooding of a poor neighborhood, it hardly rises to the level of crisis until someone rhetorically builds the repetition itself and the failure to respond to that recurrence into a surprise and a unique event. The explosion of the Challenger shuttle was unique but not because a spacecraft exploded on take-off or because human lives were lost in the pursuit of space exploration. It was unique in ways that were quite specific to that mission (civilian in space), to the United States (having not lost a human in space flight), and especially to the visibility of the explosion (televised and shown in schools). Many people had died in the global pursuit of space prior to that day, and many launches had been televised, but the uniqueness of that explosion came from a variety of elements that all came together in the rhetorical construction of the crisis.

Alternatively, sometimes the uniqueness of a crisis event can be precisely because of its repetition. The DC-10 crashes are an excellent example, as the first occurrence could be brushed aside as an accident, but as the same failures repeated twice, then three times, then a fourth, a unique event emerged: the combination of four crashes under similar circumstances due to the same failures on the same model plane. However, the uniqueness of such a recurrence of events is not because we have never seen a product repeatedly fail. In the auto industry alone one can easily reference the Ford Pinto, the Chevrolet Corvair, and the Ford Explorer as examples of commonality of perceived product flaws or failures that were associated with accidents. Such crises are unique, hence, not in the broader world but are egoistically unique. As Ulmer, Sellnow, and Seeger put it, "crises are unique moments in the history of organizations" (2006, 5). To put a slightly finer point on it, we might reword this a bit and say that crises are events that come as a surprise because the person or organization for whom this is a crisis has not believed the event to be probable. The distinction here is that surprise and uniqueness are not so much conditions found in the event itself but in the relationship of the event to the organizations and persons who experience that event as a crisis. A crisis event may have been (and sometimes is) entirely consistent with the history of an organization and even wholly predictable, but it comes to constitute a crisis because those in the organization had not perceived it to be probable.

Crisis as Judgment

Probably one of the most persistent elements of all definitions of crisis is the role of judgment. Definitions of the ancient Greek term *krisis*, from which the modern term *crisis* was inherited, always explicitly include judgment and decision, but nearly all the modern English definitions likewise reflect the centrality of judgment to crisis. A crisis, then, can be considered a moment in which making a good judgment is decisive or of heightened importance. In crisis communication literature judgment is about how to respond to an event so as to maximize the self-interest of the organization for which this event has been constructed as a crisis. As the crisis is understood through egoism, the judgment is likewise conducted egoistically. What does the company or organization need to do for its own survival and self-interest? Crisis communication, as a body of scholarship and as a practice, does not usually entertain the idea that perhaps the diminution or even elimination of the organization in crisis should be the preferred outcome. Why isn't the bankruptcy and dismantling of Murray Mining a potentially valid resolution to the Crandall Canyon Mine collapse? The answer is, simply, that it is a potentially valid response for many of the parties involved, but certainly not for Murray Mining. If the crisis communication is conducted egoistically by the people at Murray Mining, then this possibility is never considered acceptable. Thus, crisis judgment as an egoistic judgment takes as its basic value presumption the survival and self-interest of the organization in crisis.

Of course, disasters also require judgment, but such judgments differ considerably in what they ask. Part of the exacerbation of the damage caused by Hurricane Katrina may have been produced by applying crisis principles to a context where we expected less self-interest and a response more indicative of disaster communication. In Katrina the kind of egoism common to crisis communication approaches was made clear by its incongruity to the broader conditions and expectations. When Mayor Ray Nagin and Governor Kathleen Blanco were involved in the processes of saving face and preserving the reputations of their offices, their organizations, and themselves, they contributed to confusion and perceptions of poor response. Such face-saving strategies are common and sometimes even advisable approaches in crisis literature but inappropriate to a disaster. Katrina itself was not a crisis event and could not be rhetorically constructed as such. Katrina was a disaster. It was only after and out of the failures in judgment of those charged with disaster response that a crisis event was constructed, which overlapped the disaster but was not synonymous with it. Indeed, it was the plain egoism of elected officials and government offices that generated the crisis, for in responding to the disaster from a crisis orientation they made their selfishness obvious to all those watching. Such an error doubled the importance and type of judgment required, as one had both to find adequate disaster response to save the lives of the people affected by Katrina and to make judgments about how to handle the crisis of legitimacy and faith that the previous errors produced. If crisis thinking and its inherent egoism had never been a part of the response to the hurricane, there may never have been any crisis at all. Instead, we might have seen a more appropriate disaster response and better disaster communication. Perhaps the most important moment of judgment was made when officials decided to handle the disaster through a crisis lens and to communicate following crisis principles, a decision that would ultimately create a crisis in the midst of a disaster, compounding the tragedy of the latter.

Crisis as Heightened Importance

The importance of judgment comes from the fact that a crisis is not only a surprise that requires judgment but is of heightened importance. In medicine and pathology, crisis (and its Greek predecessor) can refer to the moment in a disease when the patient may either begin the road toward recovery or toward death. Most often this is referred to as the turning point of the disease. In crisis communication and organizational crisis studies the patient is almost always narrowly defined to the organization (usually a corporation), whose interests and survival are made paramount. The medical analogy and the quality of a moment in which hangs life or death can certainly forgive a certain egoism. We rarely fault people for making decisions to stay alive. However, this analogy overstates what is at stake in most crisis communication events. Even if these risks are related to events in which people lost their lives, crisis communication tends to focus on how best to handle the risk to the organization, in which the people outside the organization are important primarily to the extent that their plight has salience for the organization's success. What is at stake in most crisis communication is market share or a product line or perhaps the continuation of a corporation or organization. However, the crisis perspective can amplify egoism at a time when the greatest threat is actually to those outside the organization. Without being exceedingly moralistic, I think it reasonable to ask how much the continuation of a corporation, the preservation of profit, or the integrity of a brand is worth to the broader social good.

Nonetheless, even if the medical analogy can amplify the problems of the crisis orientation, the perception of an event as a key turning point or vital moment is key to the constitution of a crisis. As a definitive moment in the future of the organization and also because it is unique, a crisis is an exceptional event in which old rules and ways of thinking may not fit. Seeger, Sellnow, and Ulmer write, "during crisis establish routines, relationships, norms, and belief systems break down or no longer function" (2003, 8). This not only complicates the responsibilities and methods of judgment, but it can also unshackle egoism from ethics and allow organizational survival to overtake the organizational *ethos*. Understood as a moment of great peril to and even the potential death of the organization, when approached egoistically a crisis may allow one to invoke an extraordinary action that would set aside the normal responsibilities or constraints that mitigate egoism. Alternatively, it might be possible that in breaking down the normal modes of thinking, an organization could set aside egoism and even set aside the crisis orientation, instead stepping forward as a member of a community that faces a collective disaster.

In the latter cases, what we see is less about good crisis communication and more about good disaster response and disaster communication. Quite plainly, if an event can shake an organization from its regular daily egoistic pursuits so that it is truly willing to respond to that event without its own self-interest being dominant, it ceases to be a crisis for that organization. Even if those events still threaten the very continuation of that organization, the mode of response that subordinates self-interest to a position no higher than the interests of all the others affected by the event brings about an understanding of the event in terms of disaster, tragedy, or threat for a broader community of persons and undertakes a response that is guided by the welfare of that broader community, even if such a response might mean the end of the organization. This distinction displays perhaps the most fundamental turning point in every crisis communication event: that moment at which egoism takes the helm and guides the response away from the disaster with its broader impacts and toward crisis with its more narrow egoistic outcomes.

Crisis as Multi-Causal

One element that this re-definition of crisis does not include is the cause or origin of a crisis. This is because, as many recent cases have demonstrated, crises usually have complex and multiple causes. Even if an event can be isolated to have a specific direct cause (such as a faulty O-ring in the shuttle Challenger) the construction of a crisis from that event requires much more than just that cause. In the case of the shuttle Challenger, there was the failure to communicate inside the agency and with private contractors, additional unprecedented failures of rockets shortly before and after the Challenger explosion, the fact that the explosion was heavily televised, that day's weather conditions, and the presence of a civilian on board. These are not merely features surrounding the events or contexts for the crisis, but when we understand that O-ring failure as a crisis for NASA, all of these elements were contributory causes of that explosion becoming and being constituted as a crisis. They are as much, if not even more so, causes of the crisis event as they are dimensions of it.

Additionally, an erroneous distinction sometimes is made between crises and disasters based on their causes. Crises are considered to be caused by negligence or willful actions, whereas disasters are more commonly ascribed to nature or "acts of God." There is some limited validity to this distinction, but we must be careful not to fall into the egoistic trap of limiting responsibility to only that for which one might be a direct cause. It may be true that one cannot pin a natural disaster on a corporation or government agency, so such natural phenomena only become a crisis event to the extent that there are additional causes that shape the discourse about the disaster into a crisis for organizations that failed to respond, should have predicted, or could have prevented the negative impacts. However, disaster literature rarely makes causation a definitive part of what constitutes a disaster. The National Research Council Committee on Disaster Research states that disasters may be "natural, technological, or willful in origin" and certainly does not preclude negligence or incompetence from potential causes of disasters (2006, 15). This, however, could be taken to mean that crisis is merely a subset of disaster, and this would not be accurate. Instead, what the crisis literature demonstrates, as I have argued in this section, is that crisis is a very different orientation toward the world than disaster. One is not a subset of the other. Even if a disaster and a crisis may share specific instances and events, the disaster is epistemologically, phenomenologically, and ethically an entirely different thing, one that takes a broader orientation and does not display the narrow egoism that is definitional of a crisis.

EPISTEMOLOGICAL BLINDERS IN CRISIS STUDIES

While the depiction of crisis communication given thus far is not flattering, the most significant problem with the crisis orientation may be that it combines this egoism with epistemological blinders that make it unlikely to meet even its narrow and self-serving objectives. The epistemological errors common to the crisis orientation may explain why so many organizations concerned about potential crises have sought to build crisis response plans for a number of scenarios, only to find a crisis plan's usefulness inversely related to the significance of an event. In the first place, crisis theorists posit that if one can hypothesize the scenario and plan for it in advance, then one can either eliminate events from constituting a crisis by removing the surprise or, at minimum, provide mechanisms by which the organization's established belief systems and norms can persist through a crisis event. Such planning, while it may have some value for low-impact minor crises, proto-crises, or pseudo-crises, have generally proven less useful than expected. In his detailed study of crisis events, Francis Marra found that, in most cases, crisis communication plans were not effective. In part, this is because, he wrote, "managers treat crisis communication plans as a simple and singular solution to a crisis" (Marra 1998, 472). In short, they take the plan as an off-the-shelf response that absolves them of the need to consider the current crisis as unique and allows them to abdicate responsibility for making new judgments. Even "excellent communication plans," Marra found, "had little effect on the outcome of these crises" (1998, 473). Given the qualities that constitute crises, this seems not only unsurprising but even what we should expect to see as the results of ready-made crisis plans.

Taking the definition of crisis provided in the first section, we can see a number of explanations for the general failure of crisis communication plans. First, the quality of a crisis as a surprise forecloses the possibility that a pregiven plan will sufficiently fit the crisis. Second, a combination of a culture of egoism and a series of biases in social scientific research methods help to promote the fallacy that surprises can be predicted. Third, normative and ideological biases tend to project conditions into future predictions that are exactly the kinds of conditions disrupted by crises. Taking these elements into combination, we can conclude that a large part of what the history of crisis studies and crisis communication teaches us is that we will have inevitable interruptions in all our systems of making sense and making judgments, consistently forcing us into conditions of crisis or disaster (depending upon one's degree of egoism) that require we not only make quick judgments, but also that we find entirely new ways of making those judgments.

Phenomenology of Surprise

What is a surprise? Perhaps most simply we can describe a surprise as the experience of an event that was unexpected. Lloyd J. Dumas, a former consultant to the Los Alamos National Laboratory and professor of economics and public policy at the University of Texas at Dallas, once wrote, "Given enough time and enough opportunity, anything that is possible will happen" (1999, 283). A surprise is the experience of a phenomenal reminder of the truth of that fact. Being surprised is being confronted with the simple truth that, as Dumas put it, "events that are extremely unlikely do happen from time to time" (1999, 283). Moments of serendipity, of amazing coincidence, of tragic accident, and of disaster, can all be such manifestations. The first DC-10 crash caused by simultaneous loss of engine power and slat control no doubt came as a surprise, as the engineers had calculated the likelihood of such an event at one in one billion (Dumas 1999, 288–289). That highly improbable event then recurred three more times, which proved not only surprising but had been wholly unthinkable. These kinds of events, which surpass the simply unlikely and enter the realm of the practically impossible, happen with some frequency. They constitute surprises that require we rethink the assumptions upon which we had previously based our belief in their impossibility. These are the kinds of surprises that can be major disasters or crises and are interruptions of our established mechanisms for making sense of the world.

Take, for example, the nearly 25 percent single-day drop in the Dow Jones Industrial Average on October 19, 1987, known as "Black Monday." This was an event without precedent, without a model, and without a predictor. It was a singular event that was not simply unlikely, but was essentially unthinkable prior to its occurrence. The next-largest single-day decline for the Dow had been 13 percent on October 24, 1929 (Black Thursday). John C. Bogle, founder of the Vanguard Group and one of the world's premiere investment analysts, calculated, using accepted models of economics and the data available prior to that date, the odds of that 25 percent drop at 1 in 10^{50} (2008, 38). To drive home how unthinkable that event was, that equates to a 0 .000000000000000000000000000000000000000000001 percent chance or one in one hundred quindecillion. The number is so incomprehensible that even comparative expressions make it hard to explain. If something has a one in one hundred quindecillion chance of happening every single second, then it would happen, on average, roughly once every 3,168,808,781,402, 895,000,000,000,000,000,000,000,000 millenia.[2] Our sun's total life is estimated at roughly 10,000,000 millenia (about half of which is over). In short, one might as well call it impossible. Yet, that is exactly what did happen on that day. That event, wholly unprecedented in the history of economics, also occurred in a monitored and regulated system, and arguably the one about which we have more data and predictive work than any other.

We might be tempted to say that such surprises are so infrequent that they cannot be significant to the overall health or behavior of a system or organization. While such surprises are, by definition, the kinds of outliers that traditional social scientific and economic methods tend to dismiss, these outliers are often exactly the events most responsible for major changes and large fluctuations in systems and organizations. The outliers are, in fact, highly important because their dramatically higher impact swamps their rarity. Bogle calculates that of the 57 years of the S&P 500 index, from 1950 to 2007, the top 40 days of gains (roughly 0.3 percent of the days) account for 70 percent of the current value of the index (2008, 33). This is why, when studying anything that has to do with human behavior, be it economics, organizations, crises, or communication, Monte Carlo simulations and models based on standard distributions consistently fail. Laws of probability are often inappropriate to human behavior because, to paraphrase Bogle, the fact that an event has never before happened in human interactions and organizations is no reason to believe that it cannot happen in the future (2008, 30).

Nassim Taleb, distinguished professor of risk engineering at the New York University Polytechnic Institute and co-director of the Decision Research Laboratory at the London Business School, calls these kinds of events "black swans" and argues that not only are they the most important events in economics but also are the kinds of events most responsible for major transitions in organizations, politics, history, and a range of human social phenomena. Taleb identifies three qualities of a black swan: "First, it is an outlier, as it lies outside the realm of regular expectations, because nothing in the past can convincingly point to its possibility. Second, it carries an extreme impact. Third, in spite of its outlier status, human nature makes us concoct explanations for its occurrence *after* the fact, making it explainable and predictable" (2007a, xvii-xviii). While we might expect greater regularity and more stable distribution for the physical world, such as tossing two six-sided dice at a craps table, the world of human interactions, such as markets, politics, and history, refuse to follow such stabilities.

Retrospective Distortion

The third element of Taleb's black swans is that after an unprecedented event occurs we concoct ways to explain it as predictable. This is indicative of a methodological bias in most crisis and risk literature that frequently produces errors and frustration. Taleb refers to this as retrospective distortion, the production of an illusion of predictability by analyzing an event without accounting for the passage of time (2007a, 12–13). Such distortion is a very specific form of anachronism, in which the crisis or risk analyst looks back at an event such as the 1987 stock market collapse and, given the fact of its occurrence, builds a schema by which one could have predicted it. In short, the effort is something like reverse engineering a crisis from the occurrence of the event backwards to the things that one might have taken as indicators, if one had been looking for that event to occur. The problem in doing such reverse engineering is distortion of previous signs that will now be taken as potential indicators of the now-past crisis event by framing those signs in relation to an event that previously had not been considered possible. In short, the passage of time has taken something that was beyond improbable, such as a 1 in 100 quindecillion chance or the quadruple occurrence of a 1 in a billion event, and made it an actual fact. Given the 100 percent likelihood of the event occurring, it becomes easy to rationalize post-hoc all the signs one should have seen, but these exercises do nothing to diminish the fact that the event, prior to its occurrence, was largely inconceivable. One cannot look for the signs of something that is beyond one's imagination and, even if one can imagine something, if it is held to be practically impossible, then one cannot be expected to see anything but the most blatant and irrefutable signs of it approaching.

Compounding this generic problem of predicting outlier events is what Taleb terms the "ludic fallacy," which imposes a mathematical model of randomness onto events where we do not have sufficient data to do so. The term "ludic" refers to playing games of chance, such as craps and blackjack, where you have determined variables, predictable odds, and all possible outcomes are known in advance. Setting aside any interference from cheats or other distortions, we expect stable odds for the toss of two six-sided dice and can calculate payoffs based upon those odds. Likewise, these distributions produce rare events at the outliers but at a stable and predictable frequency. Certainly given enough rolls of two dice one will see 12 come up twice in a row, but we know the odds of such are very low. Such models depend upon Gaussian (normalized) distributions that are common in some physical environments but cannot be presumed when dealing with human interactions. In addition, such models presume to contain all possible outcomes. One cannot roll 79 on two standard six-sided dice. The ludic fallacy in crisis literature (and common to most social sciences) thus contains three key limitations when applied as a model for things like economics or other areas of human behavior: first, it proposes a stable set of odds for any given occurrence; second, it relies upon the presumption of Gaussian (normalized) distributions in populations where such distributions are not appropriate; and, third, it can only analyze foreseen events or known risks at the expense of missing the highest impact unprecedented events. Any method based upon the ludic fallacy cannot be applied to events that can have indeterminately changing odds, have irregular distributions, or in which there are unknown possible outcomes. Unfortunately, those conditions describe economics, politics, history, communication, and essentially all endeavors in which human behavior is involved, at least insofar as we are concerned with the very high-impact, very low-probability events that most concern crisis studies.

The ludic fallacy, however, has been a dominant model in the social sciences for many decades. As Taleb wrote, "several hundred thousand students in business schools and social science departments from Singapore to Urbana-Champaign, as well as people in the business world, continue to study 'scientific' methods, all grounded in the Gaussian, all embedded in the ludic fallacy" (2007a, 275). As a result, the knowledge that dominates these fields tends to have little utility when dealing with the most important types of crises: black swans. However, the criticisms that Taleb and others levy against social science do not mean that the social sciences have no value. Quite to the contrary, for low-impact events that have enough frequency to be studied with knowledge of their likely distribution and with some confidence in the range of likely results, they do just fine. They are well suited to the study of everyday minor occurrences, the general average movement and fluctuation of the markets, and similar commonalities. What they cannot formulate, by the very nature of their methods, however, are the unique events, the high-impact anomalies and outliers. Ludic models are simply not appropriate to generating odds or predictions about unprecedented events, unknown outcomes, and that 0.3 percent of the days that accounts for 70 percent of the S&P index value. We need the traditional social sciences to map the other 99.7 percent of the days that account for 30 percent of the value, and we need them to analyze the regularities of daily events and to study voter behavior in most scenarios, but we cannot rely upon them for insight or understanding into black swans without committing the ludic fallacy and misleading ourselves.

Colonizing the Future

Of course, the great risk of the ludic fallacy does not lie in misunderstanding the past but instead in projecting those misunderstandings onto the future. There can be many justifications given for committing retrospective distortion, including laying blame for political advantage, but the most common may be the prediction and prevention of future reoccurrence of such events. In some cases this may, in fact, be quite productive. Looking back can indeed help us understand certain weaknesses or vulnerabilities that we can amend to forestall a particular range of possible scenarios, but such retrospection and analysis cannot achieve the goal of domesticating the future. The goal of prediction is to make the future a property of present knowledge, but what the history of black swans teaches us is that the future will always exceed our capacity for prediction. The future will always hold the wholly unexpected and the entirely unprecedented, and while these events may be extremely rare, they also may be the moments of most significant change and challenge.

Forecasting and prediction are, of course, useful, but they also can produce a false sense of security that only amplifies one's vulnerability to surprise and the impact of negative black swan events. Sohail Inayatullah, professor at the Graduate Institute of Futures Studies, Tamkang University, Taiwan, notes that a great deal of future studies and predictive work "simply reinscribes the present even while it 'predicts' the future" (1990, 115). The plainest cause of this problem is that prediction begins from the existing conditions of the present and cannot do away with those presumptions wholesale while still retaining moorings for its predictive work. That is to say, certain elements of the present become naturalized in predictive work, which is fine for mundane predictions of minor events that do not constitute potential paradigm shifts but makes for a serious distortion when a crisis promises to change elements such as "theoretical assumptions about what is foreseeable, data assumptions about what is observable, and value assumptions about what is preferred" (Inayatullah 1990, 116). As a result, the present tends to colonize the future with current ideologies and practices. An extension of some stabilities into the future is required if one wishes to conduct any predictive analyses and is quite appropriate to low-impact, high-frequency events, but to presume these stabilities will endure makes one exceedingly vulnerable to and unprepared for their interruption.

Inevitable Interruptions

In simple systems, particularly artificial and controlled systems such as laboratories, simulations, and roulette wheels, interruptions are kept to a minimum. In more complex systems, such as ecologies, organizations, and communities, nearly everything about the system is potentially open for change. In such complicated systems it is impossible to foresee all of the ways that change might occur. As Dumas demonstrated in his studies of nuclear power and regulation of new technologies, complex systems impede our predictive abilities. Given that predictions rely upon the continuation of certain components of the present to serve as the stable variables of analysis, the interruption of those variables poses a significant problem for attempts at predicting anything but especially for predicting the range of possible events and responses in a high-impact, low-probability crisis or disaster.

The problem that faces crisis analysts and crisis communicators is that the most significant crises are exactly those in which some significant element of the established methods of making sense is interrupted. The crises that planners are most concerned with are exactly those that are rare, for, as Taleb has noted, "the severity of the event here will be inversely proportional to its expected frequency" (2007b, 198). As frequency declines, so does predictability. Indeed, the rarest events (first-time occurrences) not only exceed prediction but also often exceed what we imagine to be possible. This is not simply a matter of refining our predictive abilities so that we might account for paradigm shifts or large-impact unprecedented crises but instead is a recognition that we will inevitably find ourselves surprised by an unexpected occurrence that interrupts some of our fundamental assumptions and practices. We cannot know what the conditions or qualities of that coming surprise might be, but we should fully expect it to occur. As Bogle put it, "some surprising event out there, far beyond our expectations, will surely come to pass, an event that will carry an extreme impact, and one for which, once it happens, we'll quickly concoct an explanation as to why it was so predictable after all" (2008, 39). In short, we can be certain that the future will bring unpredictable and unprecedented events of great significance, which will require that we deploy new ways of making sense and new ways of responding.

So, even if one accepts the egoistic orientation, we are faced with a serious epistemological dilemma. For egoism to achieve its goals, it seeks to make predictions about crises that threaten the individual or organization, as well as predictions about its best responses in those crises. Yet, the most significant and highest-impact crises are both unpredictable and require that one be able to set aside presumptions about continuities and valuations. The first of these is merely a problem of the limits of knowledge and our methods of prediction. The latter, however, conflicts with egoism's presumption of self-preservation. An obsession with the survival of the self (whether that self be individual or corporate) will promote a vision of the future in which the self is not only existent but also in which the future self looks much like the present self. Thus, egoism may condition responses to crises in such a way as to limit the flexibility of individuals and organizations to redefine themselves and almost certainly will exacerbate retrospective distortion. The conclusion we should draw from this analysis is that the crisis orientation not only is ill-equipped to handle high-impact crises, but it is an active hindrance to the adaptive, spontaneous, and radical thinking that such crises may require. This latter conclusion still requires, however, a more thorough analysis of the relationship between egoism, crisis response, and good organizational response.

EGOISM AS CORPORATE CULTURE

The current dominance of egoism in crisis theory and practice may not be essential to the history and etymology of crisis but may have as much to do with a body of business regulations and practices that have developed over the past 200 years in the umbrella of corporate law. As the corporate business model has spread to nonprofits, government agencies, and higher education, it has taken its peculiar form of egoistic thinking with it. As such, the dominance of egoism in organizational culture (and perhaps American culture more generally) has made us especially vulnerable to low-probability high-impact events. While there are strategies for the subversion of egoism built into the very operation of egoism itself, those approaches risk communities and individuals being caught in the conflagration that often ensues from egoism's own self-immolation. Instead, we might remedy the current egoistic crisis orientation by adopting a more broadly engaged and democratic organizational culture and embracing the practice of imagining the impossible.

Organizational Culture as Crisis Predictor

The culture of an organization and its varying degrees of egoism or other pathologies may be possible predictors of both coming crises and how an institution will respond to such crises. Given the problems of prediction discussed in the previous section we should not overstate the role of organizational culture, but there is sufficient evidence to raise concern that egoistic organizations are especially vulnerable to some forms of crisis and that such organizations may deal with crises less effectively than those that can set aside their egoism.

Organizational communication and crisis studies scholars have already made the connection between leadership ethics and organizational crises. Ulmer, Sellnow, and Seeger have noted that the "broadest and most inclusive subcategory of intentional crises is unethical leadership" (2006, 11). This is not to say that the unethical leadership of an organization is in itself the crisis (although it could be) but that the largest category of intentional crises emerge due to unethical leadership practices or decisions. In this case the standards of ethics may range from common contemporary morality to codified organizational standards to actual violations of law, but in almost every case the crisis is a result of an organizational culture that enables or even requires unethical practices, and those practices are nearly always excesses of egoism that have put personal or organizational gain above community, organizational, or legal standards.

However, in addition to the causal relationship that often occurs between organizational culture and intentional crises, the organizational culture is also a key factor in how well an organization or individual responds to a crisis or disaster. First, as crises develop an organization often receives greater scrutiny and is thrust into the light of media attention and public inquiry. Simultaneously the organization is required to make critical decisions during crises, usually quite publicly, and often with less than ideal information. The combination of these factors may reveal a great deal about an organization, both to its members and to those outside the organization who have interest in the crisis. As Seeger, Sellnow, and Ullmer put it, "crisis often makes the organization's fundamental ethical stance highly visible" (2003, 219). In such moments strategies of organizational image management are hardly adequate. If the organization in crisis does not have a sound and ethical organizational culture and good relations with the other parties involved in the crisis, one should fully expect public relations and crisis communication campaigns to be treated with incredulity.

Finally, the culture of the organization itself will constrain or enable the organization's internal communication, its flexibility in adapting to the crisis, and its ability to set egoism aside and make decisions based on the true range of possible outcomes and broad interests and needs of the full range of participants in the event. As Marra's study of crisis plans found, "communication culture and autonomy are much better predictors of how well an organization manages a crisis" (1998, 472) than is crisis planning. Unfortunately, the crisis orientation may well draw out the worst in an organization's culture and ethics. As a central feature of the crisis orientation is the egoistic definition and valuation of events, and crisis gives that moment the metaphoric status of a fight for survival, the entire crisis approach may fundamentally be a detriment to enabling an ideal organizational response.

Egoism of Corporate Culture

Worse yet, in some organizations the problem of egoism may be so deeply embedded in their culture and structure as to make them pathological at their core. In fact, that may be the very definition of the modern corporation. This was the conclusion that Joel Balkan, professor of law at the University of British Columbia, drew after working with interviews and case studies across the corporate world. As he lays out in his introduction, Balkan found that "the corporation's legally defined mandate is to pursue, relentlessly and without exception, its own self-interest. . . . The corporation is a pathological institution" (2004, 1–2). His conclusion was not far from the reason why Justice Louis Brandeis wrote in 1933 that corporations had become "Frankenstein monsters" cobbled together from bits and pieces of case and statutory law in such a way that they were given both extraordinary legal protections and also no restraints except for profit and self-preservation.

If Balkan, Brandeis, and similar critics of corporate law and culture are correct, then we should not be surprised to see corporations and other organizations that take the corporation as their model (which may include many non-profits and universities) to repeatedly display the grossest forms of egoism. When such an organization encounters a crisis, the full weight and force of egoistic errors may come to the fore. As such, rather than actually attempting to behave as citizens and community members, corporations and corporate-style organizations look out for themselves alone, making the most useful and likely approaches to crisis communication little more than, as William Benoit (1997) has called it, "image repair discourse." The difference between an actual corporation's egoism and the pathology of those institutions that mimic corporatism in their cultures and styles is that for the corporation egoism and selfishness are written into their very definition and *raison d'etre*. They exist to return profits to investors and have a legally binding obligation (a primary fiduciary responsibility) to do so. This is why Balkan and similar critics point out that the problems with the egoistic pathology of corporations are actually imbedded in the nature of their legal existence and would require significant legal reform to alter.

The Promise of a Sociopath

Having spent a few years in exactly such a corporate culture, I think it wise to pause for a brief moment and consider that this rampant egoism does also offer a certain hope. I have no desire to make a defense of Rand or of egoism, *per se*, but I do think that one can consider the simplicity of egoism likewise an opportunity for its subversion. Absent the kind of broad legal and cultural reform Balkan calls for (which seems unlikely but nonetheless worth considering), then we might ask in what way the egoistic drive can be put to non-egoistic uses. Certainly the desire for a positive public image and good public relations can drive organizations, even highly egoistic ones, to do things that are boons to communities and the planet as a whole. However, such practices are taken up because the external pressures upon the organization either directly pose a threat to its own self-interests or have come to be understood as a likely future threat. As Sherelle Salaver, a student at the UC Riverside Sloan Center for Internet Retailing aptly put it, "how long will sociopath tendencies be able to stay profitable?" (Sherelle Salaver Blog, comment posted January 6, 2009). This adept and interesting question was bundled in a larger conversation about how corporations might change the public perception that they are uncaring and selfish entities. In the end, though, such discussions of ethics in corporate practices resort to appeals to the corporation's own self-interest (i.e., profitability).

Thus, when one deals with a purely egoistic entity (be that a person or an organization), at least one knows where the fulcrum points are upon which one might leverage some persuasion. In my own experience, I found that I could get nearly any policy or practice changed, blocked, or implemented if I could ground and evidence a compelling argument in the corporation's own self-interest (i.e., profit), especially if credit for that increase in profit could be given to the person making the decision. That single uniform decision criterion for every action meant those most skilled at grounding arguments in that goal and marshaling evidence for them would almost always carry the day. It also meant that the fear of serious financial damage due to exposure or sanction had genuine motivating force in a way that it might not in more idealist orientations. There is, thus, a clear strategic opportunity for turning egoism toward altruism and restraining the excesses of egoistic organization through severe and strictly enforced sanctions.

However, this will do little to allow such organizations to more effectively handle and cope with crises. Exceptional moments allow egoistic organizations to excuse themselves from obligations or restrictions that might normally reign in their selfishness. When their livelihood or survival is threatened, organizations and individuals alike may be more open to high-risk and ethically dubious tactics. Likewise, the orientation and structure of crisis literature and consulting often encourages organizations and persons to experience events egoistically, which increases both the potential damage done to communities and to that organization itself. Thus, the greatest promise of egoism is that it may be self-defeating. The more egoistic an organization's orientation, the poorer its response to crises and disasters, and the more likely the organization will perish or take great losses. Thus, egoism poses a dilemma that can only be escaped by setting egoism aside. To be egoistic is to look out for oneself first and foremost, but for organizations to do so imperils themselves and makes them more vulnerable to a disaster or crisis. Thus, egoism begets tragedies and disasters and exacerbates their impacts. Hence, the best way for an organization to look out for itself and its investors is to divest itself of egoism and move beyond the pathology of the corporation. At best egoism is a paradox. At worst it is form of self-immolation that catches countless people in the ensuing conflagration.

Hope and Preparedness

Such a theoretical salve to organizational egoism is, however, of little comfort to the countless people who become collateral damage in poorly handled crises, excesses of egoism, and bungled disasters. Likewise, there is little reason to believe that most corporate-styled organizations will reduce their egoistic orientations. Given these rather dire diagnoses, what advice can one offer to those who are mired in events and responses that are commonly approached as crises? Perhaps all I can offer is a bit of hope found in the disaster communication and disaster response literature along with a repositioning of what we mean by being prepared, grounded in the critiques of prediction.

When discussing the standards for good disaster response and disaster communication, the National Research Council Committee on Disaster Research noted that "people need to be confident that some person or group that shares their interests and values is well informed and is representing those positions competently in the political system" (2006, 74). If we apply this same logic from disaster communication and disaster response to crisis communication, then we might advise organizations that when a crisis emerges, the wisest response is to first step back from the egoistic orientation and look at the event from a broader community or even global perspective and through the lens of disaster. In doing so, one should see that if an organization as a member of a broader community wants to help that community weather the storm successfully it will have to build a genuine coalition with other stakeholders and surrender a significant amount of control to ensure that everyone has representatives involved in the crisis response that they trust and believe share their positions.

Preparing for such a moment cannot be done with ready-made plans but instead requires a cultural disposition be present in the organization that values the broader community and the social role of the organization. Such a disposition should lead to an amenable integration of the organization with its surroundings, both social and ecological. If organizations experiencing a crisis (or disaster) need broad coalitions of community members and stakeholders to help guide them through difficult times, but such coalitions are treated with suspicion by community members when built within crisis events, then one vital part of preparation for disasters (or crises) is to integrate such a coalition into the daily operations and general culture of the organization. In short, an organization should seek to engage not only its own members and employees but also groups within its geographic communities and broader stakeholders who have an interest in the organization's behavior. These must be relationships that build mutual trust and investment in the purposes of the organization so that when surprise events of heightened importance do occur, everyone already has representatives at the table whom they trust and who share their values or interests.

Such prebuilt coalitions of stakeholders and community members must have an actual voice in the organization's policies and behaviors if such trust is to be established. Thus, the coalition approach to preparation will require a certain level of democratization of the organization. That democratization requires inclusion of a broad range of stakeholders, giving such stakeholders actual force in the decisions of the organization, and also demands sufficient transparency as to allow stakeholders to make informed decisions and engage in substantive argument about the issues that face that organization and the broader communities in which it is a part. Such democratization and engagement will not be successful, however, if it is conducted merely as a fig leaf of public consent. Organizations that wish to both avoid the recurring self-generation of crises and be better equipped to respond to disasters must, as Marra noted, develop a superior organizational culture and *ethos*. As such, the democratic and engaged organization is an organization that incorporates its relationships and engagements with its broader communities into every element of its daily practices and makes its broader understanding of responsibility a core part of every level of its decision making and identity. In short, any organization that wishes to best avoid and be best prepared to respond to disasters (or crises) will need to shed the egoistic orientation.

While the development of broad coalitions and a democratized and community-based organizational culture will help any organization avoid and respond to disasters far better than developing crisis response plans, there is a value in using crisis scenarios as tools for training and preparation. Scenarios and simulations do help us understand and work with the kinds of judgments and conflicts that might emerge within a surprise moment of heightened importance. In other words, they train and attune our capacity for judgment. Understood in this way, the purpose of a scenario or simulation is not to decide what one would do in this specific situation but instead to help the stakeholders and community members build systems of communication and judgment that will aid them in improving responses to events that cannot be foreseen. Likewise, looking back at past events can help us to understand what approaches to judgment or methods of communication were deployed that exacerbated or ameliorated those disasters. Wisdom often comes by looking to the past so as to see how our judgments exacerbated or ameliorated a situation. As G. W. F. Hegel (1986) put it, the owl of Minerva flies only at the end of the day when the battle is done. Such retrospection can do much not only to hone one's capacity for judgment but when conducted collectively as part of the broad democratized organization will also help to build the trust and establish the relationships required when disasters strike.

Such practices, however, cannot have their full utility if practiced only as response to the repetition of the past. Given the capacities of retrospective distortion and the inverse relationship between an event's predictability and its impact, we must work to keep judgment limber by forcing ourselves to consider the impossible. Just as the 1987 single-day stock market crash was practically impossible by the standards of economists, so are many of the major disasters (or crises) that have struck communities and organizations. Our practiced scenarios should include the impossible and the absurd. Only by thrusting into our consideration the interruption of our ways of making sense can we develop the kind of adaptability required by true black swan events. It may seem odd or even inconceivable to imagine the impossible. Indeed, this was Alice's objection in *Through the Looking Glass* when the White Queen claimed to be over 100 (this shortly before turning into a sheep). Lewis Carroll, a mathematician and logician by trade (under his birth name of Charles Dodgson), reminded us through his tales of Alice not only of the possibility of imagining the impossible but the deep value in doing so. As the White Queen said, our difficulty in such imagination lies largely in our lack of practice. While some may balk at the threshold of the not-yet-imagined scenario, those wishing to develop the judgment required to adapt and respond to true black swan events will leap into the impossible with both feet. After all, when a truly surprising event turns our worlds upside down, having been upturned before is likely the best preparation.

CONCLUSION

This essay has sought to redefine and reorient our current studies of crisis. Through an examination of common crisis case studies and the discourse of crisis communication scholars and consultants, the first section constructed a definition for crisis as it currently operates in the literature: *the egoistic experience of a constructed event that comes as a surprise and poses a moment of judgment perceived to be of heightened importance.* This definition reflects the many biases and assumptions of current crisis studies, most important of which being the dominance of egoism as a valuative structure and hermeneutic. That egoism is itself both disabling to good crisis response and also negatively impacts countless people and communities caught in the disasters and tragedies that crisis orientations treat egoistically. Even more worrisome is the combination of this egoistic orientation with the epistemological blinders common to crisis studies. A combination of retrospective distortion, the ludic fallacy, and predictive overconfidence generate a reliance upon ready-made crisis response plans and image restoration strategies. Both of these approaches tend to be of little use in crisis and generally fail precisely because they both emanate from an egoistic orientation and depend on Gaussian models for human behavior.

Unfortunately these problems tend to be deeply embedded in many organizational cultures and perhaps even the very structure of American corporate law. While reform of current statutes and regulations governing incorporation are beyond the scope of this project, it does appear that if corporations are incapable of divesting themselves of egoism and truly enfranchising broader communities of stakeholders not only will such organizations hold a high risk of wreaking devastation upon those around them and, eventually, themselves, but they also will by their structure and design be unable to cultivate the kinds of judgment necessitated by surprise events of heightened importance, such as crises or disasters. If any contemporary organization wishes to be able to survive the inevitable black swan events that are to come, it must adopt two strategies that are fundamentally at odds with both egoism and the general epistemology of our current crisis literature. First, long before any surprise event occurs it must develop broad coalitions and engaged relations with its communities, which requires a degree of transparency and democratization not found in most organizations. Second, it will have to take up the cultivation of good judgment through scenarios and exercises that bring those broad coalitions together to collectively discuss how it might respond not only to the possible and historical events but also to the impossible and the absurd. To the extent that it can take the advice of the White Queen and practice imagining the impossible together, it will produce a limber and adept judgment as well as a means of collective decision-making more likely to see it through catastrophic unexpected events.

Thus, in the final analysis, the paradox of current crisis studies is that so long as we are mired in the egoism and epistemology of crisis thinking we make ourselves vulnerable to crises, while casting off egoism and thinking more broadly with a coalition of stakeholders and community members (an approach more often found in disaster studies) actually provides a better likelihood of being ready and able to weather the inevitable and yet unpredictable storm that is to come. In short, to think crisis is a fallacy that dooms us to crises. In an odd turn, our current age has made crisis thinking the enemy of judgment, and yet only well-trained and adaptive judgment will be limber enough to see us through whatever rocky straits lie ahead.

REFERENCES

Adubato, Steve. 2008. *What Were They Thinking? Crisis Communication: The Good, the Bad, and the Totally Clueless*. New Brunswick, NJ: Rutgers University Press.

Balkan, Joel. 2004. *The Corporation: The Pathological Pursuit of Profit and Power*. New York: Free Press.

Benoit, William L. "Image repair discourse and crisis communication." *Public Relations Review* 23 (1997): 177–186.

Bogle, John C. "Black Monday and Black Swans." *Financial Analysts Journal* 64 (2008): 30–40.

Brandeis, Louis. 1933. *Ligget Co. v. Lee*, 288 U.S. 517 (dissenting).

Dougherty, Devon. 1992. *Crisis Communication: What Every Executive Needs to Know*. New York: Walker and Company.

Dumas, Lloyd J. 1999. *Lethal Arrogance: Human Fallibility and Dangerous Technologies*. New York: St. Martins.

Heath, Robert L. and H. Dan O'Hair, eds. 2009. *Handbook of Risk and Crisis Communication*. New York: Routledge.

Hegel, Georg Wilhelm Friedrich. 1896. *Philosophy of Right*. Trans. S. W. Dyde. London: George Bell and Sons.

Inayatullah, Sohail. "Deconstructing and Reconstructing the Future: Predictive, Cultural and Critical Epistemologies." *Futures* 22 (1990): 115–141.

Kirkpatrick, Jerry. 1994. *In Defense of Advertising: Arguments from Reason, Ethical Egoism, and Laissez-Faire Capitalism*. Westport, CT: Quorum.

Marra, Francis J. "Crisis Communication Plans: Poor Predictors of Excellent Crisis Public Relations." *Public Relations Review* 24 (1998): 461–474.

Millar, Dan P. and Robert L. Heath, eds. 2004. *Responding to Crisis: A Rhetorical Approach to Crisis Communication*. Mahwah, NJ: Lawrence Earlbaum.

National Research Council, Committee on Disaster Research in the Social Sciences. 2006. *Facing Hazards and Disasters: Understanding Human Dimensions*. Washington, DC: National Academy Press.

Rand, Ayn. 1961. *The Virtue of Selfishness*. New York: Signet.

Salaver, Sherelle. "Is Your Company a Sociopath?" UCR Sloan Center for Internet Retailing Blog. Posted January 6, 2009. Last retrieved April 5, 2010. h0p.in/2z. Cached at h0p.in/30

Seeger, Matthew W., Timothy L. Sellnow, and Robert R. Ulmer. 2003. *Communication and Organizational Crisis*. Westport, CT: Praeger Publishers.

Taleb, Nasssim Nicholas. 2007a. *The Black Swan: The Impact of the Highly Improbable*. New York: Random House.

——. 2007b. "Black swans and the domain of statistics." *The American Statistician* 61: 198–200.

Ulmer, Robert Ray, Timothy L. Sellnow, and Matthew Wayne Seeger. 2006. *Effective Crisis Communication: Moving From Crisis to Opportunity*. Thousand Oaks, CA: Sage Publications, Inc.

NOTES

1. The texts reviewed in developing the definition of crisis and forming the critique of the crisis orientation found in this essay were Adubato; Dougherty; Heath and O'Hair; Millar and Heath; Seeger, Sellnow, and Ulmer; and Ulmer, Sellnow, and Seeger.

2. Thanks to Daniel Israel for double-checking my math in this section. These calculations account for one leap day every four years but not for the rules governing leap days in 100th years or 400th years.

Communication Ethics as Janus at the Gates

Responding to Postmodernity and the Normativity of Crisis

Ronald C. Arnett
Duquesne University

Alasdair MacIntyre, in his classic work on ethics, *After Virtue* (1981), suggested that this historical moment finds clarity in the fact that we are in a constant moral crisis, unable to agree upon what narrative and virtue structures will and should structure and guide our ethical decision making; his description offers a shorthand definition of the postmodern condition. For MacIntyre, the root cause of our moral crisis is "emotivism," defined as ethical decision making propelled by personal inclination and preference, void of any external standard or authority (1981, 11–12). MacIntyre's 1981 book, penned more than a quarter of a century ago, is no longer a scholarly revelation; the book is now more accurately understood as a brilliant prognostication of what has become *the* normative approach to communication ethics in our time. MacIntyre, one of the premier ethics philosophers, forecast the pervasive influence of "emotivism," a personal ethics decision-making tool that currently embodies a high degree of unreflective "street" credibility (1981, 129). The dominance of emotivism in this society is not surprising if one reflects upon the warnings of de Tocqueville in *Democracy in America* (1835/2004). Tocqueville coined the term "individualism" and contended against its brazen impulse to stand above history and render judgments driven by singular, personal, and monadic insight. We continue to live in the heyday of individualism with its enduring legacy carried forth in decision making governed by emotivism. Extrapolating from the work of Arendt, MacIntyre, and *Communication Ethics Literacy: Dialogue and Difference*, a work I co-authored with

Fritz and Bell (2009), this essay contends that the combination of individualism and emotivism has crippled the moral landscape associated with modernity and its sacred trinity of progress, efficiency, and individual autonomy (Arnett, Fritz and Holba 2007).

INTRODUCTION

This essay forges conceptual ground contrary to both modernity and a traditional culture, outlining a communication ethic that assumes a gate-keeping function that defies both the emotivism of modernity (with the locus of ethics inside the person) and an unreflective traditional culture (with the locus of ethics in the taken-for-granted mores of the people). The critical nature of this essay, however, is not evenly dispersed; the brunt of critique falls squarely upon the shoulders of modernity.

This treatise offers an alternative to the modern taken-for-granted assumptions of individualism and emotivism, framing a particular view of communication ethics as a gatekeeper of two postmodern goods: 1) difference and 2) the demand to unite knowing, doing, and responsibility. This communication ethics venture, understood as interpreting otherwise than modernity and its normative convention of individual autonomy, seeks to be responsive to difference and resistant to previous formulations of ethics associated with the premises of modernity and unreflective traditional society. This interpretive effort presupposes two guiding questions. First, "What is the role of an alternative conception of communication ethics in a historical moment driven by emotivism and narrative contention?" Second, "How can communication ethics reclaim a gatekeeping role in concert with a postmodern affirmation of difference without falling prey to the siren song of modernity and its attraction to individual autonomy played out by ethical monads?"

The following sections outline a postmodern understanding of communication ethics, using the imagery of the Roman god Janus—"Janus at the gates." The first section of this essay situates communication ethics in the "mud of everyday life" (Arnett and Arneson 1999, 32). The second section outlines the nature of a postmodern consciousness related to communication ethics. The third section further describes the importance of postmodernity by describing the limits of modernity through a critical portrait painted by the scholarly insights of Hannah Arendt. The final section examines gatekeeping functions of communication ethics in a postmodern age with the goal of giving shape to a communication ethic that functions as Janus at the gates, interpreting otherwise than modernity.

SITUATING COMMUNICATION ETHICS—ATTENTIVENESS TO EXISTENCE

Arendt begins the preface to her classic work *The Human Condition* (1958) with thoughtful disputation of news headlines after the announcement of the Soviet Union's launching of Sputnik on October 4, 1957 (1). She countered headline implications that celebrated the human being's finally being able to escape earth, a place that had for too long imprisoned the human spirit. Arendt's anger was not with the launching or the science of Sputnik but with a portrayal of the earth as a prison requiring the concerted effort of escape.

The purveyors of these headlines unknowingly assumed the value of escape from responsibility for a given local home in an era propelled by the energy of individualism and the ideology of "progress." The headlines unreflectively affirmed the conviction that the goal of the modern person is to stand above history or, in this case, to lord it over the earth itself. Arendt's portrayal provided an alternative; she redirected our attention back to the earth, to existence that necessitates putting our feet and our hands in real soil to which we are responsible. The earth, for Arendt, was a living existential metaphor; her conception of the human condition rejected any call to escape existence. It is our commission to be responsible for the ground that supports human life. It was not the advent of one of the major scientific wonders of the post-World War II Western world circling the globe that concerned Arendt; it was the abjection of human existence that generated her considered ire.

The wisdom of Arendt has a postmodern translation into theory and practice of communication ethics. Following Arendt's lead, any effort to frame a communication ethic that attempts to stand above history and escape the problems before us by proclaiming judgments of ethical truth untainted by the demands of everyday existence and the bias of standpoint, is, at best, a contradiction in terms and could be pejoratively cast aside as a message contrary to the phenomenological reality of the human condition. Attending to the insights of Arendt, a postmodern understanding of communication ethics with a gatekeeping function meets existence and walks upon an earth composed of a multiplicity of traditions that require disclosure. Such a communication ethic lives within the soil of existence and is at odds with any theory that results in the posturing of a communicative agent as a diviner of truth attempting to walk in an ethereal dimension that bypasses the necessity of communication ethics touching real soil with full knowledge that that ground is tainted in any given place at any given time. An existential rendering of communication ethics in a postmodern age begins with awareness of the biased, tainted ground upon which the human stands upright; theories that offer a form of existential amnesia make us forgetful and inattentive to the ground that gives us the opportunity to meet others in an upright posture and are in contrast to an ethic attentive to the earth that we all must share and care for together (Levinas 1996).

The metaphors of earth, soil, ground, and existence situate this study and practice of communication ethics. Martin Buber (1947) used the phrase, "the mud of everyday life"; indeed, this understanding of communication ethics walks in the mud of everyday life, inviting us into a place where there is dirt that requires us to do labor with our hands. In this essay, I am shifting a bit from Levinas's (1979) emphasis upon the face to the metaphor of hands in hopes of emphasizing "the doing of responsibility," even though the call of responsibility in Levinas's ethics is necessarily presupposed. When labor is not present in ethics, one witnesses a resultant problematic nourishing of the home of emotivism—*the* normative communicative ethics practice of our time.

For this essay, the emphasis rests upon labor and the hands, the phenomenological importance of doing. In reflection upon my childhood in which I watched people work on their car engines without their being able to tell another "why" or exactly "what" they were going to do next, I remember them offering, instead, cryptic comments, such as, "Watch my hands if you want to learn." I am now aware of the phenomenological significance of their doing of labor with the hands; my relatives understood phenomenologically that their thinking, or "consciousness of," was in their hands, not in their heads. For those who know phenomenology, the statements of relatives working on car engines to watch their hands is not weird or uneducated; rather, on the contrary, the suggestion is profoundly accurate from a phenomenological perspective.

Husserl would have understood my relatives as calling for attentiveness to where their consciousness was busy in the doing—in this case, in the hands. In phenomenology, consciousness is "consciousness of" (Husserl 1970, 158). If I were to be a student of those automobile engine laborers and open to learning some skill myself, I needed to be "conscious of" the doing of the hands. Emphasis upon the metaphor of the hands moves communication ethics to an embedded, situated pragmatic act of construction—the doing of responsibility in the constructing of ethical dwellings with one another. What is needed to differentiate this example, which is reminiscent of a traditional culture, from postmodernity, I will address more fully later. Nevertheless, for now, it is sufficient to state that the key difference is that postmodernity adds knowing to the doing, with a traditional culture stabilized by the unreflective. Postmodernity moves our sense of doing from the unreflective to the reflective.

Hands labor and take on soil, getting dirty as we toil in renovation, building, and construction, working together in the human condition. The reality of soiled hands in the knowing and doing of responsibility situates communication ethics within the metaphor of "dirty hands." Such a metaphor takes the notion of communication ethics out of the abstract and the pristine and into the physical admission of involvement in places where there is no final assurance for all time, only the hope of temporal clarity.

"Dirty hands" moves us from the temptation of a Pontius Pilate figure who foregoes decision making, whether confronted by major public decisions or the meeting of small ethical questions in the everyday. This essay links communication ethics to the mud of everyday life—we cannot wash our hands of the consequences of our doing of responsibility in ethical decision making.

"Dirty hands" is part of the act and the accompanying knowledge related to the knowing of limits in the doing of responsibility. Such is the reason that in the essay "The rhetoric of communication ethics" for the *International Encyclopedia of Communication* (Arnett 2008) I coined the term "tainted soil" to emphasize our inability to escape bias. Laboring in tainted ground naturally leads to a greater awareness of "dirty hands." This image is a more graphic picture of the dialogic pragmatism of Hans Gadamer. Gadamer (1975) reminds us of the power and inevitability of prejudices and bias, which simultaneously open and close any interpretive gesture. The key to this interpretive engagement is the public disclosure of bias, the admission of dirty hands that connect us to the knowing and the doing of responsibility and the inevitability of consequences.

Communication ethics casts aside the assurance of the domination of irrefutable law or the opposite side of the coin, the tyranny of self-importance, leaving us with dirty hands. From the standpoint of communication, communication ethics is the labor that dirty hands must undertake in the inescapability of bias. Situating communication ethics in this postmodern historical moment requires two temporal communicative acts. First, it is pragmatically necessary to forego lament over our lack of ethical agreement. Dialogically, human beings bring bias to the table of existence, and existence in the form of the historical moment brings bias that questions us. We must meet existence; reality indicts us, and it demands our response. An existentially situated view of communication ethics assumes that we are not in control of the world; our task is to meet what is before us, beginning with existence itself. Our inability to control existence takes us from an expectation of uniformity in ethics to a public recognition of difference, which defines this historical moment.

Second, a situated communication ethic rests in a "doing" that originates in responsiveness, not in willfulness of imposition. Responsiveness attends to the monologic demands of the historical moment, calling forth our unique sense of responsiveness. When we demand of existence rather than respond to existence, reality takes on a recalcitrance, a term familiar to the insights of

Kenneth Burke (1984, 168) and in the communication discipline by Richard Thames (1999). Ironically, as we miss the demands of the historical moment, we turn attention to our own demands of existence, missing life before us, attending solely to our own expectations. Responsiveness, not demand, assumes that when we meet a communication style that we do not endorse, it is better to offer a monologic reply than to fall into the abyss of self-important condemnation of a given discourse with a pretentious utterance, such as "that is not ethical" (Arnett 2006). A monologic reply emphasizes the ground upon which one stands and from which one speaks; a pretentious utterance simply discounts the ground of another, while assuming no responsibly to announce one's own.

Third, a situated communication ethic within a postmodern moment calls us to know the good that we protect and promote (Arnett, Fritz, and Bell 2009). Unlike my relatives who said "Watch the hands," we must not only watch the hands but attend to the narrative that nourishes the doing of responsibility with the hands and gives them direction and strength. Situating communication ethics as a gatekeeper that seeks to protect and promote difference while attending to the interplay of knowing and doing of responsibility requires public admission of tainted narrative ground that gives rise to particular virtues or communication ethics practices in particular times and particular places.

Fourth, a situated communication ethic by its very nature cannot offer answers that address all the complexity of this historical moment; situatedness presupposes limits, not ultimate insight or wisdom. A communication ethic situated in a moment of difference, not shaped by the individual, must first acknowledge the bias of one's own ground and then strive to learn more—attending to an ever-increasing mosaic (Wood 2000) of interrelated implications for the study and practice of communication ethics. The study and doing of such a communication ethic begins with a desire that has no final quenching; such inquiry necessitates a commitment to the importance of intentional labor that expands our understanding of difference in a challenging and philosophically rich intercultural world of ethics.

This admission of the unity of limits and ever-expanding texts is inherent in the interpretive approach of "philosophical hermeneutics," which demands admission of bias and additionally stresses the necessity to engage new texts. Both limits and new insight shape the scope of the interpretive. The insights of "philosophical hermeneutics" through the scholarly pen of Hans Gadamer (1977) require a pragmatic modesty that acknowledges the inevitability of tainted ground, which limits and simultaneously offers helpfulness of insight. This approach assumes that our limits texture what we see and permit the addition of our voices to the human community. What keeps limits from becoming a closed ideology is the willingness to meet texts that challenge our normative conventions, requiring us to acknowledge the everyday reality of difference. This movement back and forth between limits and the meeting of difference is how communication ethics situated in a moment of narrative and virtue contention continues to answer Kant's and Arendt's call for a commitment to an "enlarged mentality" (Kant 1790/1987; Arendt 1982). Additionally, this seesaw motion of limits and the meeting of difference reminds us that temporality is a genuine reality in such a communication ethic.

Temporality takes us to particularity, which Benhabib stressed in her alternative to George Herbert Mead's "Generalized Other" (1986, 340–341). Additionally, Sandra Harding's description of standpoint (1991) reminds us of limits that open the conversation while closing it simultaneously. One of the major figures of caring and ethics in the field of communication, Julia Wood, has centered her work on these two key insights—"the particular Other" and the inevitability of "standpoint," reminding us of the temporality of our engagements. Communication ethics situated within difference of traditions(s), resting on tainted ground, engaged with temporal conviction, and demanding of knowing in the doing of responsibility, challenges ethics situated within an autonomous communicative agent. In fact, when an ethical position of an autonomous agent meets communication ethics situated in a postmodern understanding of difference, the relevance of monologue, lecture, and rebuke may be a necessary retort demanding that another attend to embeddedness in tradition(s) and the phenomenological missteps of individual autonomy (Arnett 2006, 2009).

Monologue, or what we in everyday communicative vernacular would call monologue, can be a communicative announcement of ground upon which another stands that should be and must be taken seriously by the Other. This stress on the vitality of monologue and the recognition that dialogue cannot be turned into a universal principle is framed by multiple authors. For example, in *Between Man and Man*, Martin Buber (1947) outlines the interplay of dialogue, technical dialogue, and monologue, all three being relational principles that bring human beings in communicative engagement with one another. Additionally, all three communicative principles are needed for a holistic communicative life. Paulo Freire (1970/1974) warns against the assumption that dialogue is appropriate or necessary when engaging those who are holders of hegemonic power. Monologue has its place when a given position needs to be announced with clarity and certainty to another, suggesting not only that the information is important but more importantly suggesting that the information in many cases is nonnegotiable. Of course, monologue can be abused, but the recognition of the necessity of monologue indicates that turning dialogue into a universal principle is additionally a form of communicative abuse.

Following the spirit of Arendt, perhaps the unethical rests in any ethical theory that does not attend to existence but attempts to walk or stand above the mud of everyday life. The autonomous communication ethics agent is not evil, just content in a land of "existential homelessness," (Arnett 1994) unwilling or unable to dirty the "autonomous I" in the demands of existence before us. It is not the communicative agent that generates the wrath of communication ethics situated in postmodern difference; it is the odd combination of arrogance and ignorance. Perhaps these two terms compose the Greek understanding of *hubris,* when the person takes upon oneself the role of the gods, (Melchert 2006) attempting to walk above existence or, put even more bluntly, endeavoring to walk on water without getting damp. A situated communication ethic within a postmodern consciousness begins with an awareness of Otherness. We attend to existence as "consciousness of."

A POSTMODERN CONSCIOUSNESS

Postmodernity can be misused as a scapegoat for all our problems in unsophisticated "coffee clutch" discourse that blames this construct for the crisis and strife in the culture and the world community. This reading misses the descriptive function of postmodernity, which takes us to the "thing itself" in this historical moment—difference. Postmodernity makes public the ongoing connection among differing ethics in conflict and crisis; it breathes honesty into the intellectual air of our time. The postmodern connection of crisis to ethics is a pragmatic result of the public clashing of narrative and virtue systems. What is new is the public recognition of this clashing, requiring the naming of this contentiousness within a publicly identifiable vocabulary.

Our explicit and tacit admissions of this era of difference make this historical moment unique. The functional heart of a postmodern context is difference, and this difference lends itself to the gnashing of ethical teeth. The reality of crisis emerging from the clashing of ethical positions is, of course, far from distinctive, even if one limits the clashing of ethics to the generation of crisis within the West alone. The connection between crisis and ethics is not revelatory in Western literature; it is part of historically standard literary fare.

In the West we are the inheritors of many literary documents that bear witness to one of our most treasured assumptions, the clash of ethics and resultant crises. Homer provides the drama in the *Odyssey* and the *Iliad* that takes us into one crisis after another in the clash of ethics. He reminds us of battles fought, not just with honor, but with deceit and cunning in an effort to mystify and disorient before killing and conquering. The story of Abraham and Isaac is an infamous story about a crisis in ethics that puts in perspective tension between loyalty to God and the sacredness of fecundity. Abraham must deal with the question of where to land his heart; fortunately he is saved from sacrificing his son by an angel at the very last minute. (I always wondered if there was yet another ethic possible in that story—not just to love the son or God but to give oneself up before God, unable to take the life of one's own son, resting before God with conviction that one cannot destroy that which is made in the image of God. One might sarcastically suggest that such a response seems only possible in an irreverent postmodernity moment. Yet, we can sense such stirrings and questions in the book of Job.) From the beginning of a recorded history, we have tried to make sense out of the confounding of goods that lead to personal and public crisis, from the message of a blind Greek poet to the medieval confessions of Augustine, who pens one of the first autoethnographic reflections (1960).

Later in the Renaissance, Shakespeare gives us one story after another announcing the normative character of ethics disputes and the crises they bring, etching in the minds of coming generations the image of Julius Caesar offering his infamous line, "Et tu, Brute?" (Shakespeare 1908, III.I.77) In much more recent venues and on this side of the Atlantic shore, the works of Mark Twain and Herman Melville consistently remind us of the struggles of ethics and the crises that ensue within persons and those around them. The listing of such works could go on for a scholarly lifetime with story after story reminding us that the clashing of ethics and the reality of crisis is not in the least original.

The interplay of ethics, difference, and crisis is central to what we term the West. It seems we are the product of such an ongoing view of drama, first set forth in Greek tragedy by Thespis (Melchert 2006). Thus, the task of discerning the interplay between crisis and ethics is not new; rather, it is a central feature of literature attentive to the human condition in the West.

This "lack of uniqueness" argument offered above can encourage some to move postmodernity off the intellectual page, an act expected from a modernist backlash that wants to continue the pedigree of "progress" with the next evolutionary advancement. My only agreement with this assertion rests in the increasing banality of the term; it is now used in the opinion pieces in major newspapers throughout the United States, suggesting the danger of banality and unreflective engagement. The argument of banality is enticing to me, but the fundamental point that cannot be lost is the family lineage of postmodernity; it has its roots in a pre-modern, not a modern, worldview. Contrary to what many presuppose in their initial exposure to postmodernity, this construct postmodernity has more kinship with a traditional culture than with the more recent construct, modernity. Postmodernity encompasses a public awareness of existence fueled by a multiplicity of tradition(s). Modernity, on the other hand, assumes not a multiplicity of tradition(s) but a multiplicity of autonomous agents. Multiplicity, in this case, is not *the* defining characteristic of postmodernity; the defining characteristic of postmodernity is the prepositional phrase "of tradition(s)," or perhaps one could suggest "of petite narrative(s)" or "of embedded agent(s)," which casts a shadow more in line with existential insight of Augustine, who was the subject of Arendt's (1996) dissertation, than with the more contemporary monadic aspirations of Carl Rogers, who functions as a caricature figure in the twentieth-century fascination with the individual psyche.

This particular view of communication ethics presupposes that a tradition-bound culture finds a more hospitable home in postmodernity than in modernity, particularly as tradition(s) shift from the unreflective to the reflective and from the universal to the particular. The conception of communication ethics that shapes this essay places postmodernity in closer proximity to a traditional society than to the moral *cul de sac* of modernity (Arnett 1998). Postmodernity has its unofficial origins with the original Sophists, who argued that traditions mattered but were not universal; tradition was understood by them as intimately tied to the character and nature of a given *polis*. Traditional culture from its beginnings was tradition-based, narrative-driven, and the shaper of identity framed as an embedded agent—Athenian, Spartan, Roman, etc. Understanding such a lineage connection between the premodern world and a postmodern world unmasks modernity as an aberration, a mistaken link in the chain of the human condition. Such an assertion deconstructs the linear nature of "progress," walking contrary to the insights of Hegel and Kant (1998, 1927). Postmodernity has more in common with traditional culture than it does with its immediate predecessor, modernity.

The first postmodern target was an unreflective narrative of the faith, begun by Nietzsche (1996, 1997). Heidegger takes this critique in to existence itself and goes after the autonomous individual (1962). Gadamer eventually puts a term to this line of inquiry—bias and prejudice that shapes the tainted ground upon which we stand. Foucault and Derrida continue a critique of the autonomous individual, stating clearly the connection between oppression and unstated bias of a dominant culture. Rorty calls for recognition of "petite narratives." (1991) and Levinas reminds us to listen to an "ethical echo" that is *a priori* to social construction, framing a view of agency that is derivative, not originative, in nature (Arnett 2004). We find in these authors a demythologizing of metanarrative and the autonomy of the individual, while framing a derivative understanding of human identity.

The insights of Nietzsche, Foucault, Derrida, Rorty, and Levinas confound normative expectations, rejecting traditional ethical paradigms and rendering ever-new ethical possibilities with which we must contend. They make public what oppressed persons have known since the first exclusion from a group, whether through a class or caste structure or some other characteristic a human being cannot control—we are limited and situated persons in the human condition. Their work does more than offer an ethical description—they prescribe. As Levinas contends, ethics is both a prescriptive and descriptive call to responsibility; what one sees ethically becomes a prescriptive mandate (1990).

Without prescriptive demand of ethics, there is no recognition of difference; no one has to then take the position of the Other seriously. Perhaps this is the reason that Gandhi, who believed in *satygraha* (loving opposition), not *duragraha* (stubborn persistence), also stated at times that one had to find a way to hit some on the head to get their attention (Arnett 1980, 34). Additionally, it is why Anderson and Cissna (2008) penned an essay on dialogue that counters "dismissiveness." Ethics without prescriptive demand is at the heart of Paulo Freire's contention that dialogue is simply not possible with those of inordinate positions of power. It is this prescriptive nature of ethics that has me revisiting the importance of monologue and our overuse of dialogue attached to ever-increasing expectations for discourse "progress" (Arnett 2006, 2009).

A postmodern consciousness does not satisfy a desire for absolute assurance or for unending progress but keeps before us the constant reminder that there are narratives and virtues/ethical systems with assumptions that are in stark contrast to our own. A postmodern consciousness leaves us only with existential proof that is temporal and fleeting in this moment of narrative and virtue contention.

Postmodernity is, as MacIntyre describes, the clashing of ethical perspectives, placing long-standing and new ethics positions in contention with one another. MacIntyre's titles such as *Whose Justice? Which Rationality?* (1989) and *Three Rival Versions of Moral Enquiry* (1990) offer an insightful look into our historical moment. However, for this author it is the work of Hannah Arendt that breaks the philosophical backbone of modernity. It is her work that guides the next section of this essay.

MODERNITY

A communication ethic situated in existence, unwilling to stand above the demands of everyday life and ever attentive to a postmodern consciousness of difference, requires the unity of knowing and doing in our responsibilities.

Such an ethic renders judgment that modernity was a mistaken path, largely fueled in communicative life by individualism and emotivism and unresponsive to the third, the neighbor, and authority that is tied to a public and external source outside the individual self. Modernity is an ethical and moral *cul de sac* that tried to escape the earth, the tainted soil from which we do and must make our ethical decisions. The call of modernity is simple—it is an ongoing form of communicative seduction that works to embarrass those who value tradition, ridicule those who understand guilt, and question those worried about the relational consequences of decisions, which Gilligan (1982) and those that followed would not permit.

Modernity lives by a secular sacred trinity: efficiency, progress, and individual autonomy (Arnett 1998). Of course, if one is riding the wave of success as a person or as a society such a description makes much sense. However, if one lives in the reality of a subaltern position smartly framed by postcolonial scholars, then the trinity of modernity begins to look like an agenda available only for a limited number of participants. Or if one is part of industrial America and must bear unemployment as jobs go overseas to cheaper labor and newscasts offer increasing commentary on high executive salaries, then one might sense that we are in the moment that is beginning to ask us to jettison modernity and its previously unchallenged assumptions. (Note: modernity is not the American Dream, a term coined by Tocqueville. The American Dream was *not* solely tied to progress. The American Dream is the dialectic of individual success and concern for the community.) Without the tension of individual success and concern for the community in dialectical tension, we put at risk the American Dream, which, at its best, is a dwelling where the subaltern can rise to the presidency of the United States. Modernity puts the heart of the American Dream at risk; modernity strikes at the core of this dialectical tradition.

Arendt in *Between Past and Future* (1977) stresses the damage done to a people without tradition. The destruction of tradition through efficiency, progress, and the autonomous individual is beyond dispute if you are not one of the chosen. To kill tradition, modernity turned guilt into a modern-day curse word that must be stopped due to its disrespectful tone implied toward the sacred/secular trinity of modernity. Guilt connects a human being to narrative ground; to slander guilt is to move people to the status of "existential homelessness" under the ruse of progress. One of the major points of Martin Buber (1947) that has been routinely ignored is the necessity of constructive guilt in an understanding of dialogue that begins with the assumption that we are embedded communicative agents who continue a conversation that was in process long before we entered a given dyadic exchange.

Modernity has foisted the new upon us as if it were a sacred part of the ritual of progress. Yet each of us knows of one "new" invention after another that did not succeed. Why is it that we now are interested in organic foods and that bicycle sales are climbing once again? Arendt's concern is that we counter modernity with labor that does not stand above existence or the earth; she would caution us against assuming that each "new" idea, policy, or invention is a genuine form of progress. Attending to the soil and the importance of labor questions the modern assumption that faster is always better. For instance, who today does not wonder how Europe could have continued its mass-transit advance while we let our cities deteriorate with the blessing of individual freedom that forgot to attend to the community as we courted individual autonomy?

Thoughtful labor requires doing work that necessitates making decisions, discerning what one is doing, after adhering to Aristotle's demand for deliberation and not before. To forget to be thoughtfully vigilant and attentive while doing labor can put oneself or another at risk, as anyone knows who has done physical labor.

Labor understands the necessity of vigilance necessary for engaging and meeting what postmodernity calls "difference." Heidegger (1962) discusses the importance of tools that are found, discovered, created "ready to hand" when one is responsive to the world before us that speaks to us in its difference (103). Arendt challenged modernity's discarding of one of the major tools of human survival in modernity—the pragmatic importance of differentiated space between public and private life. This was one of her major reclamation projects; she wanted to resituate the difference between public and private space as the first line of primal protection of the person in an increasingly "modern" world inattentive to tradition in the differentiating of public and private space (Arendt 1958, 22–79). In *Dialogic Confession: Bonhoeffer's Rhetoric of Responsibility*, (Arnett 2005) I make the case that it is this natural dialectic that provided Bonhoeffer with contending power against a corrupt public space.

In modernity, we collapsed public and private life into what Arendt (1958) calls the "social" (38). Such collapsing moved us closer to a therapeutic society in which we take our emotional lives into social places, no longer protecting the difference between public and private. Such a modern assumption can force us unreflectively as parents and as academics to celebrate novelties such as "Facebook" because we do not want to appear unsupportive of this "new social space." Yet I can guarantee that Arendt would not sit here and act "hip" while another generation is seduced by the corruption of modernity. The new technology is an old game that Arendt feared and fought against her entire life—the loss of a primal form of human protection, the differentiating of public and private space.

We are in an ethical environmental crisis that has engaged in a slash and burn of narrative life and its requisite differentiation of public and private space. Such destruction puts at risk the possibility of public authority in external sources that can call forth our loyalty, critique, and ethical engagement attentive to something other than our individual selves. Arendt sought the reconnection of authority to public standards. Once public, the questions of justice or what Levinas (1991) calls the neighbor and Bakhtin (1986) called the third take on public currency; such is the reason our own Kenneth Andersen (2003) stated that in ethics there is no such entity as a "free rider" (20). In Aristotelian terms, the problem lies with the excess and deficiency, with excess, in this case, being "emotivism" and deficiency being lack of "narrative guidance." Thus, communicative ethics requires us to understand the importance of knowing that we have "dirty hands" in the doing of labor in everyday life, which assumes that whatever we help build will eventually falter, fade, collapse, or cease all together; our labor is temporal, not for all time. This view of labor begins with an ethical short hand hermeneutic of suspicion—whatever decisions attempt to bypass deliberation about consequences to one's neighbor and longer-term impact on human existence fit with modernity's "McDonaldization" (Ritzer 2000) of the human condition, attempting to bypass existential indicators that tell us that toiling in the soil and renovation are ethical metaphors aligned with the unending labor before us.

Perhaps it was because of the lack of ground that de Tocqueville (2004) deemed individualism far worse than the natural attitude of selfishness. Modernity puts at risk the places we can turn to for learning and guidance, whether for noble or selfish reasons. The postmodern task is one of reclamation of ground but with reflective engagement. Our conventional communication vocabulary works to retrieve the notions of public, narrative, and rhetoric as central to ethics in such an era. The postmodern assignment is to move gatekeeping functions outside solipsistic assertions and demands by a given individual and renew our commitment to external public authorities that require our honest engagement.

Narratives function as forms of external public authority. When they are in disarray, ethical guidance in decision making falls logically to an emotive self from the hermeneutic bias of an era we call modernity. From a modern perspective the crisis in narrative life is problematic, but from a postmodern perspective this is a time of emancipation and liberation and a call for responsibility and gatekeeping. Christopher Lasch (1985) stated that modernity moved us toward being *The Minimal Self*, alone, without ground under our feet. The charge of a situated understanding of communication ethics is not to give ground to another but to remind the Other consistently of the necessity of the need to find ground through external narrative support for ethical decision making. Such is the gate keeping role of Janus at the gates.

JANUS AT THE GATES

This final section outlines the prescriptive rhetoric of Janus at the gates. The notion of "gate" is understood as a metaphorical fulcrum point, the locus of ethical decision making. The Roman god Janus offers a picture of choice and direction with heads pointed in two directions simultaneously. The theoretical and practical goal of this image of Janus at the gates is to reclaim a gatekeeping role, one that is responsive to the postmodern conviction of difference in the knowing and doing of ethics. The key is to remind ourselves of consequences and accompanying responsibilities associated with ethical choice in discourse.

The connection of postmodernity to Janus begins with the assumption that this is a god image that no one any longer worships. Criterion one for this communication ethic is movement from the assumption that a transcendent deity in the form of *deus ex machina,* which Bonhoeffer critiqued, can deliver us into sacred truth without our responsible engagement mixed with certainty and doubt (Arnett 2005) and, in addition, that no sacred secular trinity (modernity's efficiency, progress, and individual autonomy) can deliver us to a promised land either. Criterion two: This communication ethic is the recognition of the dialectic—two faces pointing in opposing directions. It is difficult to understand an ethical act that does not offer its contrary at the same time. As we attend to one person, we miss the face of another. This speaks to Levinas's (1998) complicated understanding of violence related to ethics. The final criterion is that of gatekeeping in communication ethics; Janus was the gatekeeper of beginnings and endings. This view of communication ethics personified in the image leads to the conception of an existential deity, Janus at the gates, reminding us that our choices have consequences and we must choose with care.

The picture of Janus as a Roman god is of interest and relevance to this rendering of communication ethics. First, two faces pointing in opposite directions, signifying change as one of the most enduring Roman gods who did not have conceptual beginning in Greece. Second, change—Janus represents a moment of change or birth—consider the term "January." Over time, the image of what the gates meant changed, with particular emphasis upon war and peace, but what was consistent was the focus upon the consequences of choice released by the opening of the gates. Janus is the gatekeeper between the past and the future, war and peace, beginnings and endings. Interestingly, there is no counterpart of Janus in the Greek set of gods. He was a unique and long-standing god within the Roman tradition. Janus was the gatekeeper between barbarism and civilization. In philosophy of communication terms, Janus was the god of the dialectic; he was the god that reminded us that all of life is a double-edged coin. Where there is joy, there is the real possibility of sorrow. Janus pointed in two directions simultaneously, reminding us that as we move in one direction, it is the other possibility that gives identity to the direction we have chosen.

With Janus's emphasis upon the changing of the seasons, it is a short step to temporality in decision making. Such temporal engagement calls for constant responsible participation; this theme was at the center of Dietrich Bonhoeffer's (1972) existential understanding of ethics in a "world come of age"; his insights forecast the coming of postmodernity. This postmodern moment renders life within the dialectic of narrative support and lack of undue and complete assurance; we function without an undisputed "safety net" of assurance, keeping us outside modernity's dream of the unstoppable force of progress.

Our tacit assumption in modernity is that even when ethics in crisis happens, the final outcome will somehow push us forward into a "better place" that we cannot understand at a given moment. In postmodernity, Janus at the gates, in the guise of communication ethics, must whisper to each passerby—"Choose carefully, and do not assume that 'progress' will follow you or another." We live in a time in which the Hegelian dialectic of the clashing of ethical positions no longer offers quiet solace during struggle, conflict, and the discounting of another with the assumption that such acts will ultimately open the gates of progress.

Postmodernity does not worship at the altar of tradition or the altar of the up to the minute. The altar of postmodernity is existence without a safety net, reminding us to choose carefully, discern the consequences, and never lose the wisdom that we might be wrong, and even if we are right, our best ethical efforts will not suffice forever. Janus at the gates offers us no implicit assumptions that can propel our mistakes into the field of progress, for even our victories may take us to a place far afield from an assumption called progress.

Janus at the gates reminds each passerby to engage a postmodern-situated communication ethic with the following monologic declarations: (1) choose carefully; the ethical choices have consequences; (2) in such an age, ethics finds life in learning about difference and death when used as a weapon (Arnett, Fritz, and Bell 2009) (3) narrative ground requires an ongoing commitment to learning, keeping it from falling into an ideological abyss;

(4) human identity is derivative, dependent upon existence, not the illusion of individualism; (5) today is a public reclamation project of attending to authorities and standards external to ourselves that permit questioning and argumentation that will drive us from the trenches of a private "dark ages" fueled by emotivism; (6) not only must our identity be understood as derivative, but so must our understanding of communication ethics, which is derivative of the ground, the soil, and the narratives that give rise to ethics in communicative practices; and (7) labor with dirty hands is not afraid of tainted soil, moving communication ethics from self-righteous proclamation to a pragmatic humility cast in sweat, doubt, and temporality.

This project ends with a celebratory cheer that we are recognizing that modernity is a historical moment that has been unmasked. With that unmasking has emerged the reality of difference—we live in a time of narrative and virtue contention. In such a moment, it is simply to be expected that ethics and crisis will walk hand in hand in a time of disagreement. Communication ethics is the doing of theory in the mud of everyday life. Communication ethics is not pristine. Communication ethics is not pure. Communication ethics is not predetermined. Communication ethics is the *phronesis* of everyday life; it is the communicative act of Janus, reminding us that there is no one right answer but that each path has consequences, some that we can see and others far beyond the range and depth of our vision and imagination. This view of communication ethics lives within the power of Greek tragedy, which sought no end called progress; the notion of tragedy has more kinship with limits and suggests "There is no final assurance of progress, just as there is no final protection from the possibility of human tragedy."

In the book I just completed with colleagues, *Communication Ethics Literacy: Dialogue and Difference*, we ended with a conviction that I deeply believe—this is a wonderful time to be studying communication ethics. To engage such a study in an era of difference makes us learners, not tellers; considerers, not proclaimers. We may have lost our assurance of one tradition. We may have lost the assurance of individual autonomy. But what we have gained is so much more precious—we have been welcomed back to existence, to taking care of Arendt's earth once more. Communication ethics in such a moment takes us to many traditions, able to honor the Sophists once again who ask us to give up false identities for the risk of meeting existence. We are asked to be the person who comes to the party and notices what is actually there—talking with many, engaging the party—rather than the person who comes to the party and stands aloof from the messiness and strangeness of persons that actually make the party possible.

This moment is not defined by loss but by gain. We had to lose to gain; such is the way real progress works. We lost a universally agreed-upon external authority and gained a feast—"Babette's Feast" (Arnett, Fritz, and Bell 2009, 226). This moment welcomes us back to Arendt's earth, which is no prison; it is a dwelling where we can labor with and for one another. I can hear Levinas adding softly: "It is a place where burden can make us whole in the doing of responsibility." Janus at the gates reminds all in monologic terms and then turns to each of us with hospitality and welcomes each one of us to the table of difference. Not a bad view of communication ethics—a chance to party, a chance to eat, a chance to labor, a chance to find meaning in burden, and a chance to honor our place of dwelling by laboring with our whole being in responsible engagement with our limitations and our willingness to learn from difference. Janus just asks us to acknowledge the reality of what we bring to the table—tainted ground and dirty hands. Janus at the gates is our image of a situated understanding of postmodern communication ethics, lecturing us to learn as we meet difference, welcoming our limits, and waiting for the world as it is made anew by the interplay of the imperfections we bring with our desire to learn from that and those we do not know, while walking in the mud of a place that we no longer desire to escape.

REFERENCES

Andersen, Kenneth. E. 2003. "Recovering the Civic Culture: The Imperative of Ethical Communication." Carroll C. Arnold Distinguished Lecture, National Communication Association. New York: Pearson.

Anderson, Rob., and Cissna, Kenneth. N. 2008. "Dismissiveness and Dialogic Ethics: Rush Limbaugh and Public Dialogue." In *Communication Ethics: Between Cosmopolitanism and Provinciality*, ed. Kathleen Glenister Roberts and Ronald. C. Arnett, 263–284. New York: Peter Lang.

Arendt, Hannah. 1958. *The Human Condition*. Chicago: University of Chicago Press.

——. 1977. *Between Past and Future: Eight Exercises in Political Thought*. Middlesex, England: Penguin Books.

——. 1982. *Lecture on Kant's Political Philosophy*. Chicago: University of Chicago Press.

——. 1996. *Love and Saint Augustine*. Chicago: University of Chicago Press.

Arnett, Ronald. C. 1980. *Dwell in Peace: Applying Nonviolence to Everyday Relationships*. Elgin, IL: Brethren Press.

——. 1994. "Existential homelessness: A contemporary case for dialogue." In *The Reach of Dialogue: Confirmation, Voice and Community*, ed. Rob Anderson, Kenneth N. Cissna, and Ronald. C. Arnett, 229–246. Cresskill, NJ: Hampton Press.

——. "Interpersonal Praxis: The Interplay of Religious Narrative, Historicality and Metaphor." *Journal of Communication and Religion* 21 (1998): 141–163.

——. 2004. "A dialogic ethic 'between' Buber and Levinas—A responsive ethical 'I.'" In *Dialogue: Theorizing Difference in Communication Studies*, ed. Ron

Anderson, Leslie Baxter, and Kenneth N. Cissna, 75–90. Thousand Oaks, CA: Sage.
——. 2005. *Dialogic Confession: Bonhoeffer's Rhetoric of Responsibility*. Carbondale: Southern Illinois University Press.
——. "Through a glass, darkly." *Journal of Communication and Religion* 29 (2006): 1–17.
——. 2008. "The rhetoric of communication ethics." *In International Encyclopedia of Communication*, ed. Wolfgang Donsbach. Hoboken, NJ: Wiley-Blackwell/ International Communication Association.
Arnett, Ronald. C., and Pat Arneson. 1999. *Dialogic Civility in a Cynical Age: Community, Hope, and Interpersonal Relationships*. Albany: State University of New York Press.
Arnett, Ronald. C., Janie. Harden Fritz, and Annette Holba. 2007. "The rhetorical turn to otherness: Otherwise than humanism." *Cosmos and History: The Journal of Natural and Social Philosophy 3*(1) (2007): 115–133.
Arnett, Ronald C., Janie Harden Fritz, and Leeanne Bell. 2009. *Communication Ethics Literacy: Dialogue and Difference*. Thousand Oaks, CA: Sage.
Augustine. 1960. *The Confessions of Saint Augustine*. Trans. J. K. Ryan. New York: Doubleday.
Bakhtin, Mikhail Mikhailovich.1986. *Speech Genres and Other Late Essays*, eds. Caryl Emerson and Michael Holquist. Austin: University of Texas Press.
Benhabib, Seyla. 1986. *Critique, Norm, and Utopia: A Study in the Foundations of Critical Theory*. New York: Columbia University Press.
Bohoeffer, Dietrich. 1972. *Letters and Papers from Prison*. Trans. Reginald Fuller, Frank. Clark, et al. New York: MacMillan.
Buber, Martin. 1947. *Between Man and Man*. Trans. R. G. Smith. London: Collins Clear-Type Press.
Burke, Kenneth. 1984. *Permanence and Change: An Anatomy of Purpose*. Berkeley: University of California Press.
Freire, Paulo. 1974. *Pedagogy of the Oppressed*. Trans. M. Bergman Ramos. New York: Seabury.
Gadamer, Hans Georg. 1975. *Truth and Tethod*. New York: The Seabury Press.
——. 1977. *Philosophical Hermeneutics*. Trans. David E. Linge. Berkeley: University of California Press.
Gilligan, Carol. 1982. *In a Different Voice: Psychological Theory and Women's Development*. Harvard: Harvard University Press.
Harding, Sandra. 1991. *Whose Science? Whose Knowledge? Thinking From Women's Lives*. Ithaca, NY: Cornell University Press.
Hegel, Georg Wilhelm Friedrich. 1998. *Phenomenology of Spirit*. Trans. A.V. Miller. Delhi, India: Motilal Banarsidass Press.
Heidegger, Martin. 1962. *Being and Time*. Trans. John Macquarrie and Edward Robinson. Boston: Blackwell.
Husserl, Edmund. 1970. *The Crisis of the European Sciences and Transcendental Phenomenology: An Introduction to Phenomenological Philosophy*. Evanston, IL: Northwestern University Press.

Kant, Immanuel. 1927. *Idea for a Universal History on a Cosmo-Ppolitical Plan*. Trans. Thomas de Quincey. Washington, D.C.: The Sociological Press.

———. 1987. *Critique of Judgment*. Trans. Werner S. Pluhar. Indianapolis: Hackett.

Lasch, Christopher. 1985. *The Minimal Self: Psychic Survival in Troubled Times*. New York: W. W. Norton and Company.

Levinas, Emmanual. 1979. *Totality and Infinity: An Essay on Exteriority*. Trans. Alphonso Lingis. New York: Springer.

———. 1990. *Nine Talmudic Readings*. Bloomington: Indiana University Press.

———. 1991. *Otherwise Than Being; or, Beyond Essence*. Trans. Alphonso Lingis. Dordrecht: Kluwer Academic Publishers.

———. 1996. *Proper Names*. Trans. Michael B. Smith. Stanford: Stanford University Press.

———. 1998. *Entre Nous: Thinking of the Other*. New York: Columbia University Press.

MacIntyre, Alasdair. 1981. *After Virtue: A Study in Moral Theory*. London: Duckworth Publishers.

———. 1989. *Whose Justice? Which Rationality?* Notre Dame: University of Norte Dame Press.

———. 1990. *Three Rival Versions of Moral Enquiry: Encyclopaedia, genealogy, and Tradition*. Notre Dame: University of Notre Dame Press.

Melchert, Norman. 2006. *The Great Conversation: A Historical Introduction to Philosophy*. Oxford: Oxford University Press.

Nietzsche, Fredrich W. 1996. *Human, All too Human: A Book for Free Spirits*. Trans. R. J. Hollingdale. Cambridge: Cambridge University Press.

———. 1997. *Daybreak: Thoughts on the Prejudices of Morality*. Trans. R. J. Hollingdale. Cambridge: Cambridge University Press.

Ritzer, George. 2000. *The McDonaldlization of Society*. Thousand Oaks, CA: Pine Forge Press.

Rorty, Richard. 1991. "Habermas and Lyotard on postmodernity." Vol. 1 of *Essays on Heidegger and Others: Philosophical Papers*. Cambridge: Cambridge University Press, 164–76.

Shakespeare, William. 1908. *The Complete Works of William Shakespeare*. Ed. Sidney Lee. Chattanooga, TN: Jefferson Press.

Thames, Richard H. 1999. "Nature's Physician: The Metabiology of Kenneth Burke." *Kenneth Burke and the Twenty-First Century*, ed. Bernard L. Brock. Albany: State University of New York Press, 19–34.

de Tocqueville, Alexis. 2004. *Democracy in America*. Trans. Arthur Goldhammer. New York: Library of America.

Wood, Julia. T. 2000. *Communication Mosaics: An Introduction to the Field of Communication*. Belmont, CA: Wadsworth Publishing.

INDEX

www.ingramcontent.com/pod-product-compliance
Lightning Source LLC
LaVergne TN
LVHW101447200926
848483LV00001B/83
* 9 7 8 1 6 1 1 4 7 4 4 9 7 *